Procurement in the Construction Industry

Procurement in the Construction Industry

The impact and cost of alternative market and supply processes

Will Hughes, Patricia Hillebrandt,
David Greenwood and Wisdom Kwawu

Routledge
Taylor & Francis Group
LONDON AND NEW YORK

First published 2006 by Taylor & Francis

This editon published 2013
by Routledge
2 Park Square, Milton Park, Abingdon, Oxon OX14 4RN

Simultaneously published in the USA and Canada
by Taylor & Francis
711 Third Avenue, New York, NY 10017, USA

Routledge is an imprint of the Taylor & Francis Group, an informa business

Publisher's note
This book has been prepared from camera-ready copy supplied by Will Hughes.

British Library Cataloguing in Publication Data
A catalogue record for this book is available from the British Library

Library of Congress Cataloging in Publication Data
Procurement in the construction industry : the impact and cost of alternative market and supply processes / Will Hughes ... [et. al].-- 1st ed.

p. cm.

Includes bibliographical references and indexes.

ISBN 0-415-39560-7 (hardcover : alk. paper) 1. Construction industry--Cost control. 2. Construction industry--Finance. 3. Building materials--Purchasing. 4. Contractors--Selection and appointment. I. Hughes, Will, Ph. D.

TH435.P726 2006

690.068"7--dc22 2005035157

ISBN10 0-415-39560-7
ISBN13 978-0-415-39560-1

Contents

Acknowledgements

Research of this nature is highly collaborative and could not take place were it not for the willing and enthusiastic participation of a large number of industrial partners. We are grateful to the following organizations for their whole-hearted commitment to this research: Collaboration for the Built Environment (formerly *Be*, now part of *Constructing Excellence*), who not only contributed a great deal to the management of the project and the research, but also co-ordinated the steering group for the project drawn from the following organizations:

- Amec Capital Projects Ltd
- Amey plc
- Asite
- Balfour Beatty Civil Engineering Ltd
- Building Design Partnership
- Carillion plc
- EMCOR Drake and Scull
- Gardiner and Theobald
- Gleeds
- Irvine Whitlock Ltd
- Kier Group plc
- Land Securities plc
- Marketing Works
- Waterloo Air Management
- Waterman Partnership

We should like to thank Helen Lingard and Hong Xiao, both of whom contributed significantly to the formative stages of this project, and Jenny Hong for one of the supply chain maps and associated data, prepared as part of her MSc course in Project Management at the University of Reading.

We would also like to acknowledge the large number of anonymous interviewees and questionnaire subjects who provided data for our analyses. The research project was funded by the Engineering and Physical Sciences Research Council, through the Innovative Construction Research Centre at the University of Reading, UK.

Executive summary

The UK construction industry has recently witnessed a move to innovative working practices that involve greater collaboration and partnership than has been the case in the past. While the benefits of such collaborative ways of working are widely discussed, little is known about their relative cost. Indeed, there is scant evidence of the procurement costs of even the more traditional, competitive practices. The purpose of this major piece of research was to examine whether different procurement approaches are associated with differences in procurement costs.

In seeking answers to this question, we examined the most significant procurement methods, both traditional and innovative, to identify, and where possible, quantify the commercial costs that are involved in each. The costs arise under four headings: marketing, agreeing terms, monitoring of work, and resolving disputes. Recent literature reveals that expectations vary about the impact of procurement method on procurement costs, but such expectations are largely untested. Most researchers agree that competition in construction procurement is organized wastefully, but estimates of the cost of tendering alone have varied from 0.5 to as much as 15% of construction prices. Commentators tend to suggest that the use of collaborative working would reduce tendering costs, although there has been little to substantiate this, so far. It was against this background that the research project was set up to identify and describe the procurement practices in use, and to explore, identify and measure their cost. The research benefited from the generous participation of a very active research steering group in which industry practitioners were strongly represented. Consequently, some guidance was expected on the most advantageous approaches to be adopted in the future, with particular interest being shown in the relative procurement costs of collaborative working.

For the purposes of this report *collaborative working* is defined as an approach to procurement where:

- competitive bidding is not the only criterion upon which contractors, consultants and suppliers are selected;
- some reliance is placed on the deliberate development of long-term working relationships;
- there is a limited number of interdependent participants or 'partners'.

The nature of the research has necessitated the adoption of a radical, but far more rigorous approach to the classification of procurement methods than is customary. In practice, these have usually been defined in very general terms, and a rather casual and arbitrary basis has been used to distinguish one from the other. In fact, in order to account comprehensively for the differences between procurement methods, six variables should be defined for each project. These are: the source of funding, contractor selection method, price basis, responsibility for design, responsibility for management, and extent of sub-contracting. The uncomfortable fact then emerges that, assuming at least five options under each one of the variables, there are thousands of permutations.

A wide range of research methods was applied. A comprehensive literature search was undertaken; three questionnaires relating directly to costs were administered, though only two were found to be practicable; extensive interviews were conducted with participants from all stages of the procurement process; group discussions were held with selected experienced contractors, specialists, consultants and clients; and participants at several international conferences were presented with and debated the research as it progressed. The strategy for the gathering of data was ambitious, but has resulted in the most comprehensive and exhaustive study on this topic to date.

Discussion interviews showed that companies throughout the construction supply chain are seriously engaging with collaborative working practices. There are, however, many barriers to their widespread implementation. Partnering is good for getting contractors involved at an early stage but involves serious commitment and costly negotiations to set it up. Moreover, frameworks and partnerships do not guarantee that work will actually flow from partnered clients, and the flow can be turned off at any point. This kind of risk means that few contractors and suppliers can afford to have too high a proportion of their turnover in partnering arrangements.

Some negative aspects of collaboration have been given very little attention. The development of collaborative working practices in which a limited number of contractors are required to communicate their growing body of experience and expertise to each other needs careful management to avoid the development of collusive practices. While large continuing clients have a lot of power over the markets in which they procure, there are dangers of distorting the market by enabling limited numbers of suppliers and contactors to develop monopolistic positions, if the experience of particular types of project cannot be accumulated by anyone else.

One of the major conclusions of the research, based on a survey of bidding costs, is that there is no evidence that simply the presence or absence of collaboration affects tendering costs. There is tremendous variation in tendering costs according to the extent of the work involved (for example whether it includes or excludes design), according to the participant in the process, or according to the success ratio of bids. A main contractor typically spends about 2½% of its turnover in selling its services, specialist and trade contractors about double that, and suppliers of bespoke components nearly 9% of turnover. These figures, because they are related to turnover, take into account the costs of tendering for work not obtained.

The interviews yielded much useful information on the ways in which firms manage their procurement. They find, for example, that the most effective form of marketing is "word of mouth" and, of course, satisfied clients leading to repeat business. While many clients say they prefer to negotiate with firms they can trust, when seeking tenders they often invite between three and six firms to bid, even in situations where there is a likely shortage of well-qualified bidders. At the other end of the process, disputes, especially between clients and contractors, are now rare, thanks in large part to partnering. Reforms to the legal processes have also resulted in fewer litigious episodes.

Design costs are high and, if included in the bidding process, greatly raise its cost. Because of this it is generally inappropriate to base design competitions on full designs. Design and Build contractors are often expected to submit designs and other solutions at their own risk, making bidding costs excessive. High bidding costs for PFI projects are caused by the extra requirements of producing solutions and costs for lifetime operation, over and above the usual design and construction solutions and cost calculations.

Theoretical developments in transaction cost economics hold out some hope for quantifying the costs of different procurement structures. But there is little empirical evidence in the research literature from any industrial sector. The complexities of this kind of data collection, coupled with the inadequacy of the theory for explaining construction organization, mean that this particular framework does not provide a useful basis either for explanation or for analysis of alternative forms of market structures in this sector.

The structure of supply chains affects the overall costs of procurement. Despite rising interest in supply chain management and integration, supply chains are seldom mapped and there is little evidence of research into the dimensions of real supply chains in the construction sector. Mapping supply chains showed that few supply chains in the sampled projects went below three tiers, and most stayed within two tiers of the main contractor. Removing layers of the supply chain by using integrated companies may not result in significant savings, because the costs of market transactions would be replaced by the costs of management, as well as the cost of paying for resources even when they are not needed for a particular project. Moreover, fragmented supply chains made up of large numbers of small companies helps with access to highly specialized skills, local knowledge and resources, as well as with risk distribution.

Wasteful practices in tendering include anonymity of competition, excessively long tender lists, diverse pre-qualification practices and poor quality and timing of information for bidders. The number of experienced estimating staff is often the limiting factor on how much work a contractor can take on.

The increasing incidence of early involvement of design and build contractors in collaborative working arrangements is welcomed universally. However, this may lead to reduced roles for professional consultants in the future.

Public sector procurement is more complex than private sector procurement, with good reason. Where the private sector is oriented towards generating money, the public sector's task is to spend money on public services. While not incompatible with each other, their tasks are fundamentally different, and there are limits on how far the commercialization of the public sector can be pushed.

Recommendations for industry to consider are:

- Continue to seek early involvement of contractors and suppliers. Although contractors are increasingly involved in the early stages of projects, parties further down the supply chain still find this rare.
- Develop methods for reimbursing costs for cancelled projects. This may require the setting up of a new kind of bond, or earmarking funds from which successful contractors can recover bidding costs in the event that a contract does not go ahead.
- Select contractors based on the value of what they contribute, rather than the lowest price. This may require developing different ways of calculating the price for the work, relating payment to the performance achieved by the facility, rather than to builders' work and materials.
- Do not strike off from future tender lists those contractors who are not in a position to bid for a particular project. Cover prices are not helpful and do not help to get a true market price. Contractors should be able to justify not bidding on those occasions when they are simply too busy. There is no reason why this should disadvantage them for future projects from the same client.
- Restrict the number of bidders in collaborative projects, especially where there is pre-qualification, to two or three that are sufficient for competition. Bidding costs that involve design proposals are expensive, and should not be multiplied by involving more than three bidders.
- Pre-qualification practices are complex and highly varied. Pre-qualifications can sometimes be as expensive as a full bid. The industry should operate on more standardized practices, with departures from the norm only where necessary.
- Inform bidders who else has been invited to tender. Bidders can usually find out with whom they are competing, so it is not realistic to believe that their identities can be concealed. Although there is a fear that revealing this may compromise the competition, it is already compromised. Moreover, bidders need to know that they are competing with firms who are seriously engaged in long-term development of their business, rather than short-term opportunists.
- Provide timely and informative documentation to tendering contractors. Too often, tendering periods are compressed due to late information and the quality of proposals is compromised due to incomplete information.
- Consider the roles of professional consultants when implementing changes to procurement practice. Professional inputs require experience, knowledge, judgement and impartiality. It is important not to compromise these by designing them out of the process.

This research has shown that there has been enormous progress in moving away from traditional competitive practices, but further moves in this direction may be limited by the nature of the market. Tendering costs vary a lot but are not necessarily influenced by the methods used: this is simply because there are many other factors that impact costs. Future research and development work should focus on pre-qualification practices and performance-based contracting.

1 Introduction

The idea for this research project into the costs of procurement in the construction industry was born at a time of great interest throughout the industrialized world in the development of innovative working practices in the management of the commercial processes of the construction industries. The new thinking included new methods of financing projects, such as the development of PFI (PPP); new methods of procurement and new collaborative working practices. These were sometimes introduced as separate innovations but in many cases there was a combination of two or three of the developments in one project or group of projects. One of the changes taking place was the increasing involvement of the client in the movement for change, not only in the public sector, but also by the private client organizations with long term construction programmes. The impetus for change was so great that it seemed that a new era was dawning in the commercial arrangements for contracting, replacing the largely confrontational and adversarial approaches of the past. There had been a considerable amount of discussion on the advantages of the new methods of working, both realized and anticipated, in terms of a better or cheaper product or faster delivery times. However, there was very little consideration on the costs of procurement throughout the whole construction process.

Prior to the commencement of this project there were a number of industry views about the costs of procurement, most of which concerned the costs of tendering. They were typically quoted as ½ -1% of turnover or 2-3% of bid price for PFI (PPP). However, the costs of procurement are greater than the costs of tendering alone and include marketing prior to tendering. In addition, some other costs of the project, including the costs of enforcement are affected by the type of procurement. Each member of the construction team, including the client, incurs costs in procuring any project. All these costs are of concern to the research team. Innovative approaches to business processes reduce the reliance on competitive tendering and focus instead on building co-operative and collaborative business relationships over the medium to long term. The costs in collaborative arrangements change substantially but remain significant.

The commercial process involves making deals. The costs associated with bidding and tendering are only a part of the costs of making deals. In order to get into a deal-making situation, any construction vendor (seller of construction goods and/or services, including consultants) will engage in marketing activity of some kind. This costs money. The process of striking the deal, then, is the second step in the commercial process. If the deal is struck, a contractual relationship comes into being, and since this involves services, not just goods, it is a relationship that occupies time. Thus, the negotiated deal reflects only the situation at the time the deal was struck, and may not reflect the continuously evolving relationship.

However, the parties to the contract are bound by their agreement. Their performance will be subjected to continuous or intermittent monitoring to ensure that they deliver what they have promised. This monitoring is a cost of procurement and is the third stage in the commercial process. Finally, when the work is completed, or otherwise brought to an end, the contracting parties may be in dispute about the extent to which each has fulfilled his or her part of the bargain. The resolution of such disputes is the fourth stage in the commercial process.

The selection of building contractors usually depends on some form of market competition. In those countries moving away from centrally planned economies, such as China, there is a clear perception that competitive tendering increases the quality and efficiency of contractors' performance (Wang *et al.* 1998, Shen and Song 1998). Similarly, in Hong Kong, the move towards fee-bidding for consultancy services seems to be gathering momentum (Ng, Kumaraswamy and Chow 2001). Connaughton (1994) described how to apply competitive processes to the selection of consultants, at a time when moves towards fee-bidding were growing increasingly popular.

Competitive tendering has been used extensively for a long time, but there is plenty of evidence that it does not necessarily result in value for money (see, for example, Pasquire and Collins 1996). Indeed, there are signs of increasing disenchantment with competition on price, particularly in the UK (Lingard *et al.* 1998, Wong *et al.* 2000). The costs associated with traditional tendering practices seem unnecessarily high, due to excessive detail in the information required for the bidding processes (Poh and Horner 1995).

In Sweden, Svensson (2001) examined the factors that influence the choice of consulting firms for international projects. He found that long-term relationships were at least as important as traditional skill and experience factors. Such research highlights the marketing effort that consulting firms require when obtaining work, although there is no assessment of their costs. It also has resonance with the moves in the UK towards innovative working practices, and away from straight price competition. This move follows the public sector's discarding of compulsory competitive tendering and replacement with the idea of "best value". Indeed, the use of compulsory competitive tendering for local authorities led to widespread criticism of lowest-price bidding in the UK. The recent move to "best value" as opposed to "lowest price", following the Local Government Act 1999, should help to avoid the negative effects of fragmentation and duplication in terms of monitoring, supervising and inspecting (Nettleton 2000).

While there are plenty of articles extolling the virtues and benefits of different ways of working, the benefits are rarely placed against the costs associated with collaborative working practices (see, for example, Gordon 1994, Pokora and Hastings 1995, Rahman and Kumaraswamy 2002). It is far from clear that the adoption of new ways of working is anything more than lip service, with sub-contractors continuing to report the same treatment at the hands of main contractors, regardless of the incidence of these new collaborative working practices (Greenwood 2001). Although many new ways of working are referred to as collaborative, this is a vague term which can be misleading because every construction project requires collaboration between many people at every stage. In the context of this study, it is taken to mean procurement where:

- competitive bidding is not the only criterion upon which contractors, consultants and suppliers are selected,
- some reliance is placed on the deliberate development of long-term working relationships,
- there is a limited number of partners.

Miller *et al.* (1999) report that it is unlikely that collaborative working methods will produce promised gains and reduce transaction costs if the sub-contractors are not fully integrated into the process. And there is plenty of evidence that they are not. Indeed, Moore and Dainty (2001) demonstrate that there are enormous cultural barriers in typical professional practices in construction that prevent the achievement of team integration in novel procurement routes. The idea that there is a single better way to organize construction projects seems not to be borne out by empirical work (Kumaraswamy and Dissanayaka 1998). Lingard *et al.* (1998), in their review of the literature on this topic, concluded that there was much still to be done in evaluating the impact of different ways of working on the costs of entering into a contract. Wang and Wu (2000), in their development of "cyberspace" procurement methods, acknowledge the enormous cultural difficulties in re-engineering the tendering process such that it could be automated.

The need for this research is clear. There has been tremendous interest from academics and industrial partners in equal measure. This report sheds light on some very complicated and difficult issues connected with contemporary moves towards different ways of working. It deals with the costs of procurement to all of the parties involved with the construction process and points out some of the implications for the economy as a whole. This should assist in evaluating the balance between costs and benefits of various different procurement routes.

2 Review of existing knowledge

There is much research that helps provide a context for work in this area. The research was based on the idea of investigating the cost consequences of the kind of collaborative working practices called for by various strategic reports on the construction industry. In order to deal with these consequences, the distinguishing characteristics of different procurement methods need to be articulated. Further, since there are commercial transactions at every level in the supply chain, work on supply chain mapping sheds light on the scale of transactions in construction projects. There has been much theoretical work, based on the costs of transactions, which seeks to provide explanations for the existence of firms and markets (i.e. the decision about whether to make or buy supplies). While this work provides a useful starting point for defining transaction costs, the focus is not on the choice between firms and markets, but on different ways of selecting and contracting with supply chain partners. There has been some interesting work investigating the factors that affect the costs of procurement, and some work on assessing the costs. This section of the report describes the previous work done in these areas, providing the context for the research carried out.

2.1 CALLS FOR CHANGE

Tendering was among the main issues tackled by the Latham Report, a joint government-industry review of procurement and contractual arrangements, published in the UK in 1994. It would probably be no exaggeration to say that Latham sees traditional tendering as "the root of adversarial attitudes" (for example, Latham 1994: 58). In his earlier, interim report, he levelled a number of serious criticisms at the industry's traditional tendering process. These included the sheer expense of complying with tender procedures (particularly for design and build work), the excessive length of tender lists, and the existence, particularly at the level of sub-contract tendering, of "malpractices" such as "Dutch auctioning" and "bid peddling" (Latham 1993: 28). The prevalence of Dutch auctioning and bid peddling are perfect examples of the failure of traditional tendering: carried out with the aim of price-reduction, the effect of both is to undermine the willingness of a prospective contractor to commit to best price in the initial tender (Construction Industry Board 1997a: 21). Latham's recommendations on tendering show a particular concern that public sector clients, while being aware of European Union Directives, should tender selectively and adhere to established codes of procedure. Clients who "seek tenders on a design and build basis" should be particularly aware of the costs of bidding for this type of work, and modify their selection procedures accordingly (Latham 1994: 57). Latham also noted that local authorities were being "severely hampered by being forced to accept the lowest tender" often neglecting other aspects of "value for money" (Latham 1994: 58). Four of Latham's 30 specific recommendations in the executive summary (Latham 1994: vii-ix) refer to tendering:

- The Construction Industry Council should publish a code of practice dealing with "project management and tendering issues".
- "Tender list arrangements should be rationalized ... and advice issued on partnering".
- "Tenders should be evaluated ... on quality as well as price" and recommendations on tender periods should be followed.
- "A code of practice for the selection of sub-contractors should be drawn up" ... with "short tender lists" and "fair tendering procedures".

In 1995 the Construction Industry Board (CIB) was set up with the primary objective of implementing the Latham recommendations. Among the CIB's publications, several relate to tendering, and these include codes of practice for the selection of consultants, main contractors and sub-contractors, as well as related publications on partnering, briefing and pre-qualification. The common features of the codes for tendering (Construction Industry Board 1997b and 1997c) are the requirements that:

- Clear and transparent procedures should be followed.
- Tender lists should be compiled systematically and be as short as possible.
- Conditions should be the same for all tenderers.
- Confidentiality should be respected.
- Sufficient time is to be allowed for tendering.
- Sufficient information should be provided.
- Tenders should be assessed on quality as well as price.
- Tender prices should not change on an unaltered scope of works.

In July 1998 a rather more radical approach to tendering was exhibited in another UK construction industry report, commissioned by the Department of the Environment, Transport and the Regions and produced by a "task force" under the chairmanship of Sir John Egan. The report reflected a "deep concern that the industry as a whole is under-achieving" and that "too many of the industry's clients are dissatisfied with its overall performance" (Egan 1998: paragraphs 4-6). In order to achieve the ambitious performance targets set in the report, Egan observed that the industry will need to make "radical changes to the processes through which it delivers its projects" with a view to "eliminating waste and increasing value" (Egan 1998: chapter 3). The report makes specific reference to the need to "replace competitive tendering with long term relationships based on clear measurement of performance and sustained improvements in quality and efficiency" (Egan 1998:

paragraphs 67-71). This involves "new criteria for the selection of partners" based, not on "lowest price, but ultimately ... best overall value for money" (Egan 1998: chapter 4). According to Egan, "the most immediately accessible savings from alliances and partnering come from a reduced requirement for tendering". While this admittedly "goes against the grain, especially for the public sector", and causes concern with all clients that that they are getting value for money, it is considered vital, since "cut-throat price competition and inadequate profitability benefit no-one" (Egan 1998). The influence of the Egan report has prompted a number of further initiatives in the UK industry, including the *Movement for Innovation* (M4i) and the *Best Practice Programme* of the Construction Innovation and Research Management Division of the Department of the Environment, Transport and the Regions, both of which retain an interest in the reform of tendering practice.

However, perhaps the single most significant shift in procurement policy has come from the implementation, on 1 April 2000, of the Local Government Act 1999. Under the Act, the requirement for compulsory competitive tendering had been abolished, and replaced with "Best Value Procurement". Broadly speaking "Best Value" requires a council to seek improved performance by whatever means is best. The legislation requires authorities to *challenge* whether existing practices are still relevant, *consult* on better cost-effectiveness, *compare* its performance with others through benchmarking, and *compete* with the best solutions (Joseph Rowntree Foundation 1999: 47). The result is that local authority clients are enabled to experiment with alternatives to tendering.

2.2 CHARACTERISTICS OF PROCUREMENT OPTIONS

Within the construction sector, procurement has become a complex and difficult topic. This is because it refers not only to what is bought, but also to a diverse array of methods for acquiring a huge range of buildings and infrastructure facilities. Before developing a general view of how procurement options differ, it is useful to identify the main features of current procurement approaches. There are methods of contracting and/or funding, methods of selection and methods of payment.

2.2.1 Methods of contracting

- *General contracting:* Design is provided by independent consultants in direct contract with the client or (in the public sector) designers that are part of the client organization. A separate contract for the construction of the project is placed with a building contractor, who then sub-lets elements of the work. Payment for the building work is monthly, based on how much work has been done to date, in relation to a tendered lump sum, based on unit rates in a contractual bill of quantities. Selection is normally by competitive tender, though the tender list is often pre-selected, rather than open.

- *Design and build (pure):* Design and Build (D&B) is a procurement system where a single organization undertakes the responsibilities and risks for both the design and construction phases. There may be various levels of employer-involvement in the design: in the "pure" form of D&B, the client engages a building contractor at the outset (after competition or otherwise) who is then responsible both for the design and the construction of the work. The typical payment method for D&B is a lump sum, payable in monthly instalments, based on a cost document that forms part of the "Contractor's Proposals" which is itself a tendered or negotiated response to the "Employer's Requirements", documents that form the basis of the contract.
- *Novated design and build:* A widely used variation of D&B occurs when the client employs a design team for the early stages of the project (typically up to the planning permission stage) to prepare the outline design and an "Employer's Requirements" document. A building contractor is selected by tender or other means and the design team is then transferred to this builder, and it is this transfer of contracts from the client to the builder that is called novation. The design team then continues to prepare a detailed design. The effect is that much of the employer's traditional design control is retained in the early stages, while passing ultimate responsibility for the design to the contractor. Novated D&B has been criticized for restricting the commercial position of building contractors, and for creating conflicts of interest for designers, though both groups appear to tolerate the system because of its appeal to clients, and it seems to be very widespread in the UK, more so than pure D&B.
- *Management contracting (MC):* Management contracting emerged as a response to the need of developers to take more of the commercial risk on construction projects than would be the case in general contracting. Coupled with the growing trend for building contractors to sub-let all of the work, this resulted in the need to procure only a project's management and co-ordination input in conjunction with a close relationship between client and contractor. Since the aim is to ensure that a management contractor faces little financial risk for the performance of others, the management contract is usually let on a cost-reimbursable (cost plus or target-cost) basis, with a fee bid for managing the project together with an agreement for reimbursement of expenses incurred. Typically the management contractor will sub-let all of the actual construction work to "Works Contractors".
- *Construction management (CM):* The contractual arrangement and services rendered by a Construction Management firm are not dissimilar from those under Management Contracting. But relieving such an organization of contractual risk for the performance of sub-contractors is much more effective if they are not contractual intermediaries. Thus, the most significant characteristic of CM is that there is no general contractor; instead there is a series of direct contractual links between the Client and the Trade Contractors, making the role of the CM more like a consultant than a contractor. The

arrangement is used particularly by experienced clients on projects with short lead-times.

- *Package deals:* In terms of allocating risks and responsibilities to contractors, there are many ways of increasing the scope of a contractor's work. For example, many large engineering projects are structured as *Engineer, Procure and Construction*, an arrangement typically used for large projects, such as oil rigs, harbours and docks. Under this kind of arrangement, the EPC contractor takes on the responsibility for carrying out all of the design, constructing and commissioning work, such that the client only has to pay. *Turnkey* deals provide another example of package deals.
- *Systems involving service agreements:* In some cases, organizations may be contracted to provide other inputs. These could include responsibility for activities such as commissioning, operation, and maintenance. For example, *measured term contracting* may enable a client to call on a contractor to provide building work during a specific period of time, as and when required, based on pre-agreed rates for specified types of work. This is often used for ongoing maintenance contracts. Sometimes, as stand-alone agreements, such arrangements fall into the category of "facilities management", rather than "construction" contracts. In some cases, however, they can be part of a complex *design-build-operate* deal.
- *Collaborative working*: The inputs for a project come from many organizations, but there is also a variety of ways in which these organizations define their commercial relationships. There is much to be gained in developing long-term collaborative working arrangements, and for this reason, the continuity of a business relationship can have a significant impact on the way that a business transaction is carried out, including the means of selecting contractors and/or consultants. One result of this is that construction projects are less frequently perceived as one-off, discrete, contractual deals. The trend towards longer-term arrangements is clear with phenomena such as *framework agreements* and *serial or strategic partnering*, which involve long-term relationships over programmes of work rather than an individual project. Important examples of this are the private sector frameworks (for example, those operated by BAA and many of the large supermarket chains) and, in the public sector, *ProCure21* (a framework for the delivery of NHS Estates projects) and *Prime Contracting* (used by Defence Estates). The advantages lie in saving the costs of re-bidding each individual project, the prospects of continuous improvement from one project to the next, and a more predictable workflow for the supply-side. Disadvantages include the chance of relationships becoming too comfortable, and the client's loss of access to "market value" that comes with abandoning repetitive tendering. To offset these, such deals often include incentives or performance improvement regimes. In some of these longer-term agreements a competitive element is retained.

- *Private finance:* The normal arrangement for construction work is based on finance procured by the project initiator (the "Employer", "Client" or "Owner"). However recent years have seen some dramatic changes in the way many projects, particularly those in the public sector, are financed.[1] The terms used for such projects include *Design-Build-Finance-Operate* (DBFO), *Private Finance Initiative* (PFI), and *Public-Private Partnership* (PPP). The project initiator[2] starts the process by inviting outline bids from selected organizations. These vary in nature, but typically involve a purpose-made company, called a "special purpose vehicle" (SPV) which is made up of funders, contractors and operators. The list of competitors is progressively reduced at each stage as the proposals are developed and this is an important factor, as the "up front" bid preparation is purported to be very costly indeed. The successful bidder will enter an "upstream" contract with the owner, and "downstream" contracts with constructors, suppliers and service providers. In some cases (for example, in power generation) there are additional long-term contracts with the users of the service provided. The deal is ultimately concluded when responsibility for the facility is transferred back to the owner – typically after a number of years of operation.

2.2.2 Methods of selection

Whether relationships are extended or one-off, they need to formed in the first place, and this is accomplished with a greater emphasis on either competition or co-operation. There is a range of possible levels of competition, from open tendering to single negotiation. At the competitive end of the spectrum *open* and *selective tendering* rely on price as their only or main criterion. However, some clients adopt a more co-operative outlook and favour negotiation, where non-price criteria play a significant part. *Two-stage tendering* is a hybrid approach that seeks to exploit the advantages of negotiation and competition. It also accelerates the process by permitting the overlap of design and procurement. The appointment of a contractor is carried out in two stages: Stage 1 is competitive, and based on costs for preliminaries, overheads and profit; the Stage 2 appointment is made after a satisfactory open-book negotiation of the final price. The co-operation at this stage can also help bring significant value improvements, not least through the early involvement of specialist contractors. One means for parties to develop more collaborative relationships is *partnering*; the idea of basing contracts on concepts of trust and co-operation instead of price competition.

---

[1] The concept is not new: the infrastructure of several countries was created this way in the 18th and 19th centuries, and on a smaller scale builders ranging in size from large "package dealers" and "builder developers" to small "domestic" builders commonly offer finance.

[2] Variously referred to as "owner", "client" or "sponsor", and in most cases a central, local or *quasi*-governmental body.

2.2.3 Methods of payment

There are two basic approaches to paying for any service or supply, namely a price-based, or a cost-based approach. The former typically involves a price (often, though not necessarily submitted in competition) for a given *output*, while in the case of the latter, the provider is reimbursed for its *inputs* (for example, the costs of the time and materials expended). In the case of price-based payment, the risks of process inefficiency lie with its supplier, whereas the purchaser is at more risk with a cost-based system. For this reason cost-based systems are not common for construction work; they are reserved for special circumstances and specific services.

- *Price-based systems:* These include lump-sum arrangements that range from Guaranteed Maximum Price (GMP) to remeasurable contracts. A GMP is a common feature of projects where design control is exercised by the contractor; remeasurable contracts are characterized by the contract bill of quantities (or schedule of rates) and are typical of traditional general contracting, particularly in civil engineering. In remeasurable contracts, the amount of every type of work carried out is measured after it is finished, so that the price can be calculated with reference to the rates in the bills of quantity. By contrast, fully measured quantities refers to the situation where the contractor is paid for the quantity specified in the bills, not for the quantity measured on site (but any work subject to a variation order is, of course, remeasured). In theory, the quantity executed should not differ from the quantity described in the bills.
- *Cost-based systems:* Cost-based payment methods include cost-plus and target-cost. *Cost plus* contracts effectively remove the risk of variable production costs from the contractor, who is paid on the basis of time spent and materials used rather than on the basis of a tendered price. Management contracts are often carried out on this basis. *Target cost* contracts are an incentive to ensure maximum efficiency from the contractor. Although the payment is made, as before, on the basis of time and materials rather than on the basis of a tendered price, a target cost is established at the outset (usually in competition). When the out-turn cost is within the agreed target there is a sharing of the savings between the employer and the contractor; when the out-turn cost exceeds the agreed target there is a sharing of the extra cost. The precise proportions of these shares are a matter for individual negotiation.

2.2.4 Conceptual distinction between procurement methods

The range of procurement options and the different ways of combining them reveal a confusing array of methods, as each is distinguished from the other on a different basis. What is needed is a more conceptual way of distinguishing the various methods from each other, because contracting, funding, selection and payment methods can all be combined with different modes of ownership in many different

Table 1: Conceptual definitions in procurement

Category	Examples
Ownership, initiation and funding	Owner-financed, public sector-financed, developer-financed, PFI, PPP
Selection method	Negotiation, partnering, frameworks, selective competition, open competition
Price basis	Work and materials as defined by bills of quantity, whole building, a fully-maintained facility, performance
Responsibility for design	Architect, engineer, contractor, in-house design teams
Responsibility for management	Client, lead designer, principal contractor, JV
Amount of sub-contracting	0-100%

ways. Simple labels such as PFI, partnering, general contracting, and so on, communicate little about the way that a project is structured. In fact, these different methods are not mutually exclusive categories of doing work. By grouping the procurement variables, it ought to be possible to describe any construction procurement method. These aspects of procurement are variously represented in every procurement method, even though a procurement method is usually characterized by only one factor. For example, PFI involves private sector investment in major infrastructure projects, whereas in DB the contractor's responsibility for design is the defining characteristic. To bring more clarity to the distinctions between procurement methods, the categories in Table 1 are suggested. Using Table 1, it is clear that describing a project as PFI does not indicate where the responsibility for design lies. Describing a project as general contracting does not indicate how much sub-contracting there is, or how the contractor was selected. To adequately describe how something is procured, all six categories are needed.

2.3 APPROACHES TO SUPPLY CHAIN MAPPING

In identifying how clients award work, and how contractors and consultants obtain work and exploring the costs associated with different tendering approaches and contractual and non-contractual arrangements for collaboration, the focus is certainly on costs associated with tendering. However, it seems that multiple layers of sub-contracting may add a great deal to the total cost of procurement. In the past, there were only a limited number of participants in the designing, manufacturing and construction processes for any one project. However, rapid technological advancement of products and services, coupled with the use of external expertise for the supply of construction related services and products to deliver construction work (be it a specialist design service, skilled labour, manufacture or installation), have also led to a high degree of sub-contracting as part of the typical solution in the industry (Gray and Flanagan 1989). Consequently there is an ever increasing number of participants in the construction procurement process (Tombesi 1997). Although these specialists bring advantages with them, the short-term project-related relationships between these participants are increasingly becoming complicated to set up and manage. It is therefore not surprising that management of time, cost and quality is becoming increasingly difficult. As a result, building clients are in search of new relationships with the construction industry to achieve their goals.

The recommendations in various reports (such as Latham 1994 and Egan 1998) propose strategies for reducing inefficiencies and costs. Increasing product/service complexity and outsourcing have led to increasingly complex and dynamic supply chain networks. This seems to impact on the cost of procurement. The important decision in the construction industry is not whether to outsource or not, but how best to configure the relationships in the complex network of contracts that characterize UK construction projects. If supply chain practitioners and researchers are to manage and control supply chains, it is helpful to map their supply chains as a visual aid. Mapping supply chains offer many opportunities to both practitioners and researchers to gain a more comprehensive understanding of the structure of construction supply chains, the potential number of firms within a tier and its impact on procurement costs. By focusing on the key players that comprise the largest value of procurement costs, supply-chain management promises considerable impact on procurement performance.

Empirical case studies are useful in exploring the dimension of supply chains, about which little seems to be known. This section describes the benefits of mapping construction supply chains and how to describe and analyse the structure. A framework for mapping construction supply chains is developed, focusing on the structural characteristics of the selected supply chain such as the potential number of tiers and firms involved. The study indicates that the construction supply chain is neither hierarchical nor pyramidal as suggested in the mainstream supply chain literature, but a mixture of both. The mapped structure indicates there are three tiers of suppliers from the main contractor to the sole distributor. The case study's most significant finding is the significantly large number of firms involved within the first and second supplier tiers, rather than a series of tiers of firms procuring for the project or a long chain of suppliers as perceived by most construction management pundits. It also brings to attention the importance of visualizing the organizational structure for strategic planning and understanding of the supply chain.

2.3.1 Supply chain mapping process (information gathering)

Roberts (2003) and HCi Journal (2003) suggest the use of interviews to draw a flowchart or trace the flow of relationships. Questions should be related to the relationships being mapped. For example questions may include:

- To whom does each member sell?

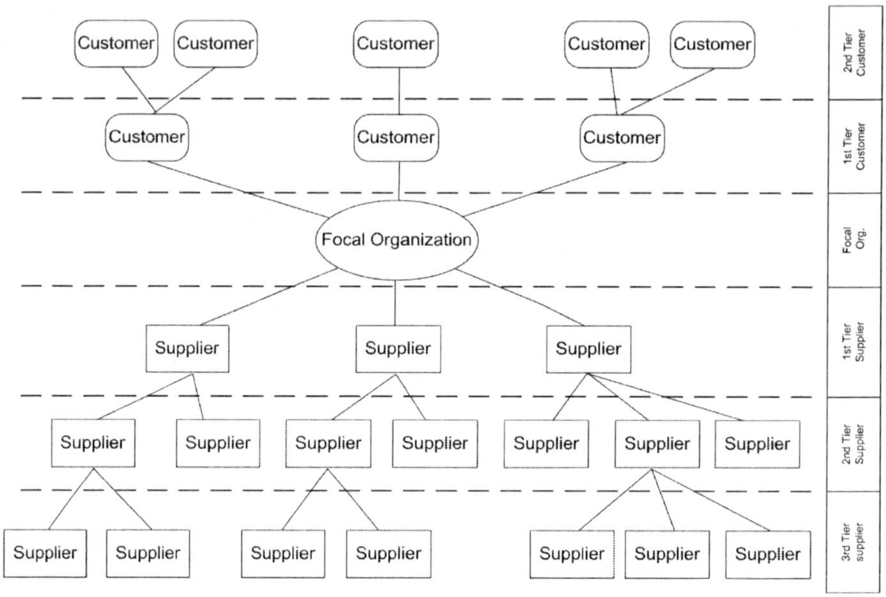

Figure 1: Supply chain model (After Fawcett and Magnan 2001, London 2000)

- From whom does each member buy?
- Who are the key partners in each chain?
- What are the functions of each partner?
- How many different functions occur along the chain?

For the information gathering process, Ljungberg (1998) suggests a combination of "real time" walk through and process construction approaches. Thus participating members can be interviewed about what happens in their supply chain. The following framework will help map out a simple supply chain (HCi Journal 2003):

- Divide a sheet of paper into as many sections as the number of tiers of suppliers and customers envisaged. Label each section in relation to the organization's section as first tier customer, second tier customer etc. or first tier supplier, second tier supplier etc. The organization's section can be labelled as the focal organization.
- Identify and note each mentioned customer or supplier in the corresponding sector or tier on the map as a labelled box. The customers who directly initiate

the focal organization's actions are placed in the first tier customer sector and those customers' actions initiate the first tier customers in the second tier customer sector and so forth. Likewise the suppliers whose actions are directly initiated by the focal organization are placed in the first tier supplier sector and those suppliers whose actions are initiated by the first tier suppliers in the second tier supplier sector and so forth.

- Connect all the labelled boxes drawn on the map with lines representing the links or relationships between the partners. An arrowhead can be used to symbolize the direction of flow of the relation being mapped.
- Alternatively, identified customers and suppliers can be asked to document their own maps using this. The organization mapping the whole chain then verifies the individual partners' maps to make sure they are all compatible and comparable to each other. The individual maps are then attached to each other at their respective interfaces.

Figure 1 shows a model derived from these steps. In describing the supply chain, Lambert *et al.*'s (1998) three critical structural dimensions can be adopted. The vertical structure refers to the number of tiers within the supply chain, which is in effect, the number of different functions that occur along the supply chain and indicates the degree of specialization. The horizontal structure will refer to the number of suppliers or customers represented within each tier. The vertical position will be the relative position of the focal organization within the end points of the supply chain.

There are, of course, some wonderful anecdotes about tendering practices, most of which, of necessity, remain anonymous. For example, in an office development in London the building already on the site had to be demolished as a first step prior to construction. In the main contract, the contractor had allowed about £80,000 for this item, and his description of the work included health and safety statements and the use of proper equipment and so on. However, this contractor sub-contracted the demolition to another, for about £40,000 and he sub-contracted it to someone else who in turn sub-contracted it again. In the end, the demolition contract was carried out over a week-end by two men with a truck, who basically pulled the building down by tying ropes to their truck and literally pulling it over piece by piece. For this, they were paid £8,000. These layers of sub-contracting appear to have cost 90% of the amount that the client paid for this work. Such anecdotes highlight the desperate need for some robust data on this issue.

2.4 THEORETICAL WORK ON THE COSTS OF TRANSACTIONS AND COSTS OF TENDERING

Transaction costs are all those costs which are not costs of production. Coase (1937) suggested that transaction costs were the determining factor as to whether a firm produces inputs in-house or buys in goods and services (make-or-buy decision). They arise from the transfer of ownership or property rights. One way of thinking about them is to envisage a Robinson Crusoe economy, where there are no other parties involved, no concept of ownership or property rights and no need or opportunity therefore to make agreements. In this case all costs are production costs. Transaction costs may arise when Man Friday comes along and some agreements have to be made. They exist whenever there is economic organization, which means that they are, in practice, universal. They include the costs of:

1. the parties to a transaction finding each other and communicating directly or indirectly;
2. the drawing up of agreements and contracts;
3. the definition and inspection of goods involved in the transactions;
4. the keeping of records;
5. enforcement of the agreements and contracts.

In the construction industry, items 1 and 2 above are very high because of the complexity of the process of producing a building or other works. The client is purchasing a product which he cannot see in advance, because it is custom-made and, when he agrees to purchase it, is not in existence. Finding the right contractor to produce the facility and agreeing a price is complex and requires binding contractual arrangements to enforce the agreements made. Many of the facilities embody a large number of different materials and manufactured goods in their fabric, which require diverse skills to put in place. Contractors generally use sub-contractors as labour rather than employing labour directly. The costs of purchasing these inputs, and ensuring that they conform to specification, are high. All these are costs of procurement, which are examined in this study. The other transaction costs (3-5 above) are not strictly costs of procurement but their nature and magnitude are influenced by the procurement methods and are therefore relevant.

2.4.1 Transaction cost economics

The analysis of the significance of transaction costs has developed into a branch of economic and management theories concerning the decision to employ factors of production directly or to sub-contract part of the production process to other firms. In both cases transaction costs arise. In the first case, for example, the transaction costs include the costs of recruitment of staff. In the second case, they are the costs of finding appropriate sub-contractors, obtaining prices, either by tendering or negotiation, and arranging contracts. In a simple situation the decision about which method to employ depends on the relative costs of the two methods.

Based on the work by Coase, Williamson (1975, 1985) has developed a substantial body of theory. The choice of whether to make or buy is dictated by the relative benefits of organizing in-house production as opposed to going to the market for supplies, referred to by Williamson (1975) as the choice between *markets* and *hierarchies*. He stressed the importance of comparative analyses of the costs of different modes of organizing productive activities (Williamson 1990). But he pointed out that reality is fuzzier than a simple choice between market or hierarchy in that markets are often characterized by networks of participants, creating regular relationships in what is sometimes referred to as a *quasi-firm* (Eccles 1981). The creation of such networks is increasingly prevalent in the construction industry, with the growing popularity of partnering, supply chain management and strategic alliancing. The ideas of Coase and Williamson have been taken up by many who study business processes (Kay 1982, Holmstrom and Tirole 1989, Pitelis 1993).

2.4.2 Transaction costs in the construction industry

Several authors have sought to explain contracting in the construction industry, which is unusually reliant on transactions for organizing work, by using a transaction cost economics framework (Reve and Levitt 1984, Winch 1989, Reve 1990, Casson 1994, Winch 1995). Empirical research is emerging to support the transaction cost concepts (Lyons 1994, 1995, Dutta and John 1995). However, Winch's (1989) suggestion that the construction industry market in the UK is a case of market failure is not based upon empirical evidence. Dietrich (1994) is critical of the ideas of using traditional transaction costs theory to explain the construction industry. The application of a neo-classical analysis of production and transaction costs to the construction industry is problematic because it ignores the inherently dynamic nature of contracting and organization problems (Dietrich 1994). Dietrich develops a framework through which the costs of transactions are re-cast in terms of the management costs associated with forming and enforcing contracts and presents this as a means for comparison with production costs. Such an approach enables transaction and organization costs to be understood as the costs of management, whether in-house or not.

In the context of innovative approaches to construction business, an important challenge is to develop supply chains without compromising competition and free trade. Monopolistic situations, and the opportunism to which they can give rise, are identified as a source of high transaction costs (Williamson 1975, 1985). A significant question is how relationships can be structured such that there is some likelihood that people along the supply chain will carry out their promises. This is particularly problematic in contractual terms because supply chain management may involve interdependencies between people who are not contractually linked (Benhaim 1997).

The emergence of partnering in the construction industry is interesting in the light of transaction cost economics. Latham's (1994) review of the UK construction industry called for the investigation of partnering and this prompted Bennett and Jayes (1998) to investigate the process of partnering in the construction sector. In other industry sectors, the development of strategic long-term relationships between firms has become known as strategic alliancing. An interesting phenomenon is that some strategic alliances result in one of the partners taking over the other (Bleeke and Ernst 1995), shifting production in-house. This could reduce transaction costs since information transfer is cheaper within a firm rather than between firms (Kay 1982, Pass *et al.* 1995).

There has been little empirical work to test the reliability of the theories in industry generally or to investigate the validity of transaction cost theory in relation to the construction industry.

2.4.3 The failure of transaction cost analysis in construction

Because of the high transaction costs in construction it might, on a first look, seem more economical to bring the inputs in-house and avoid the costs of arranging contracts for the supply of materials and labour. The theoretical approach using the concept of the internal labour market as opposed to the external labour market suggests that there should be a movement towards the internal labour market and away from sub-contracting (Buckley and Enderwick 1989). In fact, in the UK the trend has been in the opposite direction ever since the end of the Second World War.

Hillebrandt and Cannon (1990) list five characteristics of construction which affect the division of work between that carried out by the contractor directly and that which is sub-contracted:

- The finite construction period of each project.
- The wide geographical spread of location of projects and especially that of large ones which can be undertaken only by major firms.
- The uneven requirement for specific skills over the life of the project.
- The wide diversity of skills required such that any one contractor may not be able to supply all of them.
- Fluctuations in the demand for any particular type of work.

These factors far outweigh the theoretical reasons which favour the internal labour market and result in the widespread sub-contracting of the UK industry. Buckley and Enderwick accept that these factors, as well as others related to the control of the workforce, explain the situation on the ground. There may well be similar problems in applying transaction cost economics to the tendering situation. The mere fact that sub-contracting seems more expensive than direct labour is not sufficient reason to call for a reduction of sub-contracting. The important decision in the construction industry is not whether to outsource or not, but how best to structure the relationships in the complex network of contracts that typify construction projects.

Construction projects are required to be constructed where they are to be used, thus ensuring a wide geographical dispersion of projects. A large British construction firm may be operating anywhere in the UK and very often anywhere in the world. Transport costs of materials and components are such that it is economic to procure them as near to site as practicable. If the contractor had produced some of them in-house, the costs of transport would generally cancel any cost advantage in production and transaction costs. Moreover, the diversity of components and materials used in different projects would make it unlikely that more than say 10-20% of inputs could be produced in-house (Hillebrandt 1984: Table 10.5).

A similar situation arises in labour requirements. Before the Second World War and for a few years afterwards contractors employed their labour force directly. This was easier when contractors were smaller and often operated in a smaller geographical area. At that time, they would find continuous employment for their best men, but a large number of operatives, although directly employed, were in fact casual labour, taken on for a few weeks while their particular trade was required on a project. This was unavoidable because apart from wide geographical spread of work, through the course of the project any one trade was needed for only a part of the duration of the contract. The total amount of labour resource required by a contractor is small at the beginning and the end of the contract, with the bulk of the work being done in the middle. Moreover, each individual trade does not form a constant proportion of the whole (Lemessany and Clapp 1975: Table 4). If there happened to be another project nearby the workers would consider themselves lucky to be transferred. The labour-only sub-contracting system in use in the UK replaced direct, but largely casual employment.

The importance of the geographical dispersion of the products of the industry and, for any project, the variations in the employment of specialist inputs over the duration of the project completely override the arguments of transaction cost theory for in-house production and employment. There is an overwhelming advantage in the modern construction industry in sub-contracting as opposed to direct employment and of buying in materials and products rather than manufacturing. This then raises the problems of organizing the commercial process in the optimum manner, a subject on which this project makes a significant contribution.

2.4.4 Categories of costs of procurement

Following Williamson (1975), one useful starting point is to distinguish between *ex-ante* and *ex-post* transaction costs (Lingard *et al.* 1998). *Ex-ante* costs include the costs of tendering, negotiating and writing the contract while *ex-post* costs may be incurred during the execution and policing of the contract or of resolving disputes arising from the contracted work (Williamson 1975). Lingard *et al.* also consider how these costs manifested themselves in reality, and, referring to ex-*post* costs, they noted that these include direct costs such as the cost of implementing elaborate surveillance and control systems, using computer-based scheduling models and cost accounting, measuring performance, implementing a quality assurance system and providing additional layers of the managerial hierarchy

(Reve and Levitt 1984). It was noted that *ex-post* transaction costs arising from disputes and litigation could also be high. *Ex-ante* costs were tentatively identified as those incurred in (for example) contractor selection. In a later paper (Hillebrandt and Hughes 2000) the authors concentrated on developing the concept of *ex-ante* transaction costs, which they describe as "notably the selection of a contractor and determination of the price".

Lingard *et al.* (1998) have noted that some theorists, including Dahlmann (1979) argue for three categories, namely: search and information costs, bargaining and decision costs, policing and enforcement costs. These categories are roughly equivalent to the classification of transaction costs suggested by Gruneberg and Ive (2000) as: search costs, product or service specification costs, contract selection and negotiation costs, supplier selection costs, performance monitoring costs and contract enforcement costs. These are more specific than the two argued for by transaction cost economists, and for the purposes of more detailed analysis can be re-stated as:

- Pre-tendering work (such as marketing, selling, pre-qualifications and arranging framework agreements); the result is an invitation to treat.
- Tendering work (such as calculating prices, risk assessments, environmental assessments, health and safety plans and quality plans).
- Post-tendering work (such as performance monitoring, enforcement and disputes).

However, the third category is too broad in that it includes routine construction management work as well as dispute resolution. It seems both more useful and more understandable to separate dispute resolution, especially in the light of contemporary developments in procurement practice specifically designed to avoid disputes in the first place. Thus, the commercial process has been divided into four stages, the management costs of which are influenced by each other. These are shown in Table 2.

Information from one category alone would be uninformative, as variations in procedure shift costs from one category to another; for example, short-cuts made in supplier selection may result in higher performance monitoring costs. The research project is concerned with an examination of the costs associated with each of these four commercial processes in construction. Essentially the commercial process consists of four types of activity all of which involve other persons or businesses and all of which cost money.

As well as the costs incurred in selling goods and services, costs are also incurred at each stage by the buyer of goods and services. Together, these involve substantial resources which are typically dealt with as overheads, rather than individually costed. This project is the first attempt, in any industrial sector, to generate empirical data about the costs associated with finding and getting work, and the financial consequences of different approaches.

History records that all these activities were part of the construction process from early times (Hughes and Hillebrandt 2003). By the last century they were

Table 2: Stages in the commercial process

Marketing	Developing relationships and selling, including pre-qualification for preferred tender lists, forming alliances, establishing reputations.
Agreeing terms	Pricing and scoping work, estimating, bidding and/or negotiating perhaps with some element of design, and fixing a price (for consultants, defining a fee and terms of engagement); this is the process of striking the deal and at this stage a contractual relationship comes into being; the result is an offer, which may be accepted by the "customer" saying "yes".
Monitoring of work	Managing the realization of the design, monitoring performance, ensuring the carrying out of contractual obligations during the contract period; the result is the building.
Resolving disputes	Dispute resolution after the contract period, there are two types – agreeing what is owed and recovering what is owed, i.e. bad debts. Claims, enforcement and disputes; the result is the discharge of contractual obligations.

formalized, for larger projects at least. By the middle of the century Government was becoming concerned at the inefficiencies and waste which the system involved, particularly as administered by local authorities, for example, the open tendering system, and, as a result, a series of reports was produced to try to improve procurement (Murray and Langford 2003). The earlier reports concentrated on improving the tendering system. In recent times the more influential reports on the industry (Latham 1994, Egan 1998, 2002) have stressed the need for more collaboration and integration in the construction process. The Egan Report defines the results expected from such a change:

> *"The key premise behind the integrated project process is that teams of designers, constructors and suppliers work together through a series of projects, continuously developing the product and the supply chain, eliminating waste in the delivery process, innovating and learning from experience."*

It is yet early to determine how far gains have been achieved, although there is some evidence of some significant advantages. The way that collaboration takes place is still developing. Some of the processes may have very different effects in the longer term than would appear when they are set up. The idea of integrating the supply chain and the provision of a single point of responsibility between the supply chain and the client is a recent and radical change of approach to public sector procurement (Her Majesty's Treasury 1999). It was thought likely that such innovative practices would reduce the costs of tendering as well as the incidence of claims and disputes. The difference between the final project cost and the tender sum might also be lower when prices are not driven down at the outset.

2.5 EMPIRICAL WORK ON FACTORS AFFECTING THE COSTS OF PROCUREMENT

There have been some attempts to quantify the costs of transactions. Masden *et al.*'s (1991) empirical study relies on selecting a limited number variables and asking the respondents to give an ordinal score to the importance of each factor, related to 74 decisions about whether to make or buy, from one firm involved with a shipbuilding contract. These qualitative evaluations are then analysed using econometric methods to test various hypotheses about the integration decision. The limitations of this work are connected with using proxies for data instead of real cost data, and with studying only a small sample of decisions from one firm. The idiosyncrasies of the chosen industry are important, such as the complexity of the process and scheduling issues as well as the application of government regulations to defence contracts. There are too many approximations in their data for their conclusions to be reliable, even within the limited parameters of their study. They identify the difficulty of obtaining data as the key obstacle to testing transaction-cost theory. In studying the relative costs of different ways of working in the construction industry, an approach concerned with proving or disproving transaction-cost theory is not particularly helpful. While it would be good to generate data to offer insights to transaction-cost empiricists, and even data for econometric analysis, Masden *et al.* point towards the impossibility of gathering such data in any meaningful way, since every case seems to be specific to a particular situation.

There is much written about how expensive it is for contractors to bid for work. Betts (1990), for example, undertook a detailed study about the processes through which contractors go in preparing their tenders, but his aim was to help them apply information systems to the tendering process, rather than to reduce a dependency on competitive tendering. He used structured systems analysis, data flow diagrams and a data dictionary to illustrate the detailed steps of a particularly complex contractor's tender.

There is, however, evidence of efforts to apply transaction cost economics to the analysis of different procurement routes. Suraya (1997) found that client organizations simply do not have the data available for any systematic evaluation of different procurement routes. Turner and Simister (2001) have looked at the transaction costs of the client associated with setting up and administering the project; they list:

- The cost of specifying the product in the tender documentation.
- The cost of specifying the work methods (process) in the tender documentation.
- The cost of managing variations to the specification of the product during project delivery.
- The cost of managing variations in the specification of the process during project delivery.

They then consider the effect of different levels of uncertainty on the optimum contract type and find that the level of uncertainty is of significant importance.

Chang and Ive (2001) have undertaken an institutional analysis of transaction costs in construction, but their objectives are to ascertain the most effective way of organizing the market relationships. Indeed, they consciously avoid any attempts to measure directly the costs of different configurations, rejecting this approach as too cumbersome.

Becker (1993) in considering the costs associated with the standard general conditions of contract, addressed the impact that contract clauses can have on a contract price. Although the research that led to them is not explained in the paper, his findings are interesting. For example: as risk is shifted from the owner to the contractor, the contractor will increase indirect cost, contingencies and profit margins to cover the unknown conditions. There is no attempt to quantify the scale of the cost associated with the chosen contractual terms. But these findings point up some useful ideas about the consequences of different ways of doing business in construction. Most importantly, these assertions mean that any meaningful findings about the costs of tendering must also take account of the costs of contract supervision.

In relation to PFI projects, it has been noted that "traditional (contracting) arrangements primarily relied on standardized contract forms that allowed for the swift award of contracts at the expense of costly dispute resolution later in the process" (Grimsey and Graham 1997: 218), a comment that illustrates the clear inter-relationships between the various stages of a contract and the impact that economies in one stage can have on later stages. Thus, comparisons of different ways of working need also to take account of the costs of claims and disputes.

The impact of partnering was considered by Matthews *et al.* (1996), who undertook one case study in which the client, the main contractor and the sub-contractors (although not the consultants) felt that partnering would lead to lower tendering costs. Pasquire and Collins (1996) looked at the effect of competitive tendering on value. Their findings on the lengths of tender lists for traditional and design-build contracts showed that there was a lot of wasted effort in terms of abortive tendering costs for contractors, particularly in the case of design and build. From their sample, 65% of the contractors would be prepared to submit a non-competitive price (cover bid), but there is no detail on the actual incidence of cover prices. This work hints at huge costs concealed within the tendering processes, but does not attempt to quantify them.

2.6 ESTIMATES OF COSTS OF TENDERING AND OTHER COMPONENTS OF PROCUREMENT

There have been various estimates of the costs of tendering. The range of estimates is great due both to the differences in methods of procurements and also to the lack of basic information. Those contractors whose bid includes a design element have to undertake more work than traditional general contractors. How much more is not widely understood. PFI (and PPP) bids involve significant amounts of work for tenderers and substantial work by bankers and lawyers.

There is no doubt that tendering is an expensive process, which is normally absorbed as an overhead (Dawood 1995), and each bid must incorporate the cost of failed tenders (Hillebrandt 2000). For example, in 1989 it was found in discussions with contractors that they expended of the order of 0.7-1.0% of turnover in the handling of tender documentation (Flanagan and Norman 1989).

The pre-contract costs include a mass of data, such as that associated with health and safety legislation in the UK. The Consultancy Co. Ltd (1997) found that one large contractor received 5,360 pre-qualification questionnaires during 1996 which cost £589,600 to complete. The same contractor received 1,802 sets of tender documentation for which it cost £495,550 to prepare a Health and Safety response. The total of these costs during the year was £1.085m, incurred despite the fact that only 10% of the tender responses were successful. Many other researchers have pointed out the wasteful expense of competitive bidding (Pearson 1985, Dawood 1994, Pasquire and Collins 1996), but little has been done to test the assumption that contractor selection methods influence costs of the tendering process. There are many mechanisms for selecting contractors (Lingard *et al.* 1998). Each demands different types of documentation and the costs vary. Clients need to be able to make an informed judgement on the best value and not the cheapest price in their selection decisions (Egan 1998). Current practice makes such informed decisions very difficult to achieve.

Cook (1990) set out to analyse the costs to contractors of competitive bids, but his methodology involved simply asking contractors how much they spent, with no attempt to isolate the costs in any systematic way. With only 30 responses to his survey, his figures of 0.25% to 6% of turnover expended on competitive tendering are widely spread. By contrast, one research team in the UK (variously reported in Duff *et al.* 1998, Emsley *et al.* 2002) undertook extensive work to develop a neural network approach to modelling the financial impact of different procurement routes. While offering some promise in the modelling of the way that the procurement and contractor selection variables affect overall costs, they conclude that more data is needed before offering more decisive conclusions.

At a workshop involving the industrial partners for this research proposal in July 2000, it was reported that building services contractors had calculated that up to 15% of their turnover could be accounted for by "unnecessary" tendering processes, intriguingly close to the 14% associated with "organizing work" reported by Masden *et al.* (1991).

Chau and Walker (1994) published some preliminary findings that the costs of identifying and agreeing prices for sub-contracted components were cheaper than the cost of planning and monitoring the performance of direct labour. They were quite clear that the cost of the transaction was the major determining factor in whether to undertake work directly or sub-contract, which is interesting given that sub-contracting overcomes the enormous problem of how a contractor would provide continuity of work for a highly specialized and diverse workforce if it were directly employed.

Anumba and Evbuomwam (1997) highlight the high costs of tendering and mentioned calls for clients to pay tendering costs to unsuccessful bidders, but they have no suggestion as to how the cost of tendering might be quantified. Similarly, Wåhlström (1991) acknowledges the amount of work and time involved in producing tenders but does not quantify the resources. Bunn (1996) states that it is important to understand costs associated with tendering.

Interestingly, Harrison (1987) states that although it is fundamental, obtaining work is not the main objective of the estimating process. Estimating is to do with calculating the probable cost of carrying out work, whereas tendering is a separate process of deciding a price, an important distinction that frequently seems to go unnoticed in many writings on this topic. Harrison also points out that increased accuracy costs more money to achieve, and the cost rises more rapidly than the increase in accuracy. The only other thing that he says about the costs involved is that skilled estimators are scarce and expensive.

Private finance is increasingly popular with governments all over the world, as it reduces the need for them to invest capital in the short term. Grimsey and Graham (1997) estimated that in the UK, by 1997, PFI sponsors had spent more than £30m on bidding for approximately 30 schemes. Experience indicates that "[t]here has been an underestimation by all parties of the length of time to negotiate project agreements" (Grimsey and Graham 1997: 221). This amount of expenditure seems to form about 1½-3% of the total contract sums involved.

The best estimates that seem to be available for the overall costs of tendering were reported as from ½-1% of turnover for the simple costs of estimating, right up to 15% if all of the unnecessary costs associated with competitive tendering are taken into account. While these are just estimates, the principle that competition may be organized wastefully is frequently espoused in the literature (see, for example, Pasquire and Collins 1996)

3 Research

3.1 OBJECTIVES

The purpose is to identify how clients award work, and how consultants and contractors obtain work, and to explore the costs associated with different procurement approaches and contractual and non-contractual arrangements for collaboration. There are three types of cost involved: pre-tendering (marketing, forming alliances, establishing reputations), tendering (estimating, bidding, negotiating) and post-tendering (monitoring performance, enforcement of contractual obligations, dispute resolution). As well as the costs incurred in selling goods and services, costs are also incurred at each stage by the buyer of goods and services. Together, these involve substantial resources which are typically dealt with as overheads, rather than individually costed. This project is the first attempt, in any industrial sector, to generate empirical data about the costs associated with finding and getting work, and the financial consequences of different approaches. The specific objectives of this research project were defined as:

- Identify how clients award work and how consultants and contractors obtain work.
- Identify the commercial characteristics of procurement options.
- Examine any previous theoretical and/or empirical work in this field.
- Explore the structure and magnitude of the costs of the commercial process.
- Develop a mechanism for measuring the true costs of the commercial process.
- Use this new data and understanding to quantify the relationship between forms of procurement, types of project and the costs of the commercial process.
- Demonstrate how performance improvement (in terms of approaches to tendering) can be measured in practice.
- Contribute to an understanding of the most advantageous approaches to forming construction project teams.

3.2 METHOD

The research involved qualitative approaches, using individual and group interviews, to develop an understanding of the main issues involved, as well as quantitative approaches, based on questionnaires to the industry. The involvement of industrial partners in data collection, and their commitment to the project from the outset, overcame many of the usual problems that researchers have in collecting such sensitive and confidential data. By developing techniques for benchmarking the main indicators of tendering costs, the research should enable all participants in the construction industry to measure improvements in performance and to identify the most advantageous ways of forming project teams, thus increasing value for money.

The research was carried out in four stages: develop and trial data collection methods involving interviews, discussions and feasibility study of various data collection methods; two separate surveys of companies, one related to annual business activity and the other to individual bids; analysis of the structure of supply chains; in-depth interviews; analysis and synthesis of results and findings. Interviews and discussions continued throughout the process.

3.3 DISCUSSION INTERVIEWS

These interviews took place over quite a long period. They were variously intended to inform the research, review progress and find out from experienced practitioners what went on in practice. For the purposes of reporting their statements, respondents are labelled A-K, as follows:

A: Major national QS practice
B: National specialist trade contractor
C: National supplier and component manufacturer
D: Regional office of a national contractor
E: Major building and civil engineering contractor
F: Small architecture practice
G: Contractor involved in PFI schemes
H: Major national QS practice
I: Local health trust
J: Regional office of a national contractor
K: Specialist arm of a regional contractor

3.3.1 Approach and summary of interviewees

About a dozen discussion meetings were held with construction companies and professional firms. They were spread over a period of a little over a year. Some involved participation of a large number of persons from the same organization and others were one-to-one interviews. Some of the organizations were selected because they had a special way of working which seemed especially relevant to the research project. Several had membership of the Reading Construction Forum (now part of *Constructing Excellence*). Interviewees included clients, professions, contractors, trade contractors and suppliers.

The interviews generally commenced with an explanation by the research team of the objectives of the research project and the way in which it was proposed to gather information. This was normally followed by the interviewees explaining the work and structure of their organizations. Thereafter, discussion roamed over a wide range of subjects. One of the matters discussed was the design of a possible time sheet audit (described in section 3.4) to obtain data on the costs of procurement. Other proposed surveys were also considered. These discussions were very helpful to the team in finalizing the format of questionnaires.

3.3.2 Research processes and problems

One of the matters discussed was the design of a possible time sheet audit (described in section 3.4) to obtain data on the costs of procurement by the four stages in the commercial process: marketing, tendering, monitoring performance and resolution of disputes. Alternative ways of obtaining information were also discussed. These discussions were helpful to the team in finalizing the format of questionnaires.

One of the problems that emerged is that many of the professional and construction organizations do not record the time spent on marketing. One reason is that marketing is a continuing operation occurring every time a senior person in the organization meets a potential client whether socially or in the course of work. Another difficulty is that in many cases, until a vague proposal has become a project, there is not an item in the records to which the time spent on an odd meeting can be attributed. In some firms only the staff, not the partners, keep timesheets (A) or the effort may be allocated to a completely different project (B). But some try very hard to allocate costs to the appropriate contracts (B). One small firm of architects actually accounts in time sheets for the way the time of each partner or member of staff is spent in six minute increments. This firm includes in its categories continuing professional development and charitable work so that even these are properly allocated. (F)

One of the difficulties encountered in translating time spent on an activity into its cost is the confidential nature of salaries. Companies and firms are very protective of their salary structure and of individual's salaries. It might be possible to round the salaries to the nearest £5,000 or even £10,000. (E)

It was found in the discussions that there were many problems of defining exactly what each term meant (A). Very often the meaning to the research team was quite different from the interpretation put on it by the interviewee. "It would be a definition almost the size of the research in order to make one meaningful in the end" was a comment made in referring to some of the terms used. (H)

The cost of tendering for projects not won is normally put into overheads and many organizations would find it very difficult to disentangle them. (A)

The cost of supervision is another problem. How do you distinguish between "normal supervision" and the difference due to various types of procurement (A)?

In the event these problems were relatively minor compared with persuading firms to take the time and trouble actually to fill in time sheets or indeed any form of questionnaire.

3.3.3 Procurement methods

The ways in which a client or a sub-client may procure a building or other construction work are varied and complex. They are described in section 2.2.

A number of factors has to be taken into account in determining the best method for a specific project. One consultant said that he did not favour any particular procurement route. He said the most important factor was the extent of risk-sharing that the client wanted to enter into. The client would be asked to discuss his choices in a range of issues including the type of building, its complexity, the level of variation which might occur through the project, what price is required and what level of quality control. After that it would be possible to look at the appropriate form of procurement (H). It should be based on the client's circumstances and the project and never on personal preferences which are often formed by the experience of the last project which may have been completely different. Of course, any existing relationships must be considered. However, having said that the procurement process must be carefully selected, even more important is the quality of the team. If the people in the team are good enough to manage risk, he said, then the selection process doesn't matter very much. (H)

The choice of the procurement method will affect the cost of tendering. For example, if a contractor has liability for the design or part of the design, the tendering process will be more expensive. Design and build covers a wide range of different practices. Very often there is little scope for innovative design or design which will reduce the cost of the project to the client. The client often provides drawings which have been submitted for planning approval and then asks for a price on a design and build contract. In such a case the contractor simply designs the structural side and the foundations and, for example, the thickness of the hardcore in the external works. This is a form of novated design and build and is a way of shifting responsibility on to the contractor. (D)

There are "real" design and build projects where the contractor is asked to build X number of barrack places or X thousand square feet of office space. The contractor can then select all the component parts of the building from those used before and may be able to reduce the number of suppliers and sub-contractors from, say, 5,000 to 500. Standard products can be used instead of, say, designing a door with one more pane of glass than the standard or using a non-standard shade of paint. In such a situation, the number of tenderers is critical. If there are six tenderers, then some would probably send the documents back or take a cover price. If all six put in a bid, the total costs of tendering would be very high indeed. If a commitment were made never to ask more than four and to ensure that all invited would like to win the job, the contractors asked would be more interested because they would know it was a limited genuine competition. For small projects the numbers could be less.

It costs more to price a design and build project than to price any equivalent bill of quantities project. However, the costs of employing some consultants are saved and the contractor will be able to employ other consultants more cheaply than the average client because the contractor will be more willing to bargain. Moreover, the saving in the costs of construction because the contractor is in control is considerable. However, what is saved is not the cost of procurement but the direct costs of producing the building. (D)

The cost of tendering will also depend on the duration of the contractor's liability. If there is an undertaking to operate a project, e.g. DBFO or PPP, whole life costs must be estimated and the life of all parts of the works must be considered very carefully. (E)

The Private Finance Initiative (PFI) and Public-private partnership (PPP) are a type of funding, first and foremost, rather than a type of procurement. Confusion arises when one of the participants in the construction industry is involved in the financing. One of the discussions was very helpful in clarifying the nature of PFI (G). The participants are as follows:

- The client or awarding authority or concessionaire.
- The Special Purpose Vehicle (SPV), a company set up, often specifically for one project, to organize the whole.
- The co-sponsors, two or more which may include a bank and/or a contractor. The PFI project is often kept off balance sheet by working through another related company.
- The operators – probably another organization completely.
- The sub-contractors, including consultants.

The stages the project goes through are normally as outlined below:

- Investigative period when the idea is prepared and developed. All those potentially involved must be contacted and cultivated.
- Awarding authority carries out its strategic outline case (SOC).
- Government issues its credits.
- Consultants produce a hypothetical scheme.
- Issue of OJEU Notice (European Journal asking for interest).
- Pre-qualifications prepared over a period of, say, two months. Check that partners are in place and make sure it fits within the organization's plans.
- Pre-qualification documents put to awarding authority.
- After one or two months, a shortlist is announced of normally four companies and they are asked what their approach would be. Might ask for indication of cost (very difficult to produce).
- After about three months bidders down to three and Final Invitation to Negotiate (FITN) is issued.
- After about seven months a review begins with presentations and meetings with questions and clarifications.

- A preferred bidder is chosen. This is about two years after the beginning of the process. A great deal of money has been spent.
- Financial closure, i.e. successful bidder gets back what it spent so far plus a return on investment for a large project. It might be say £10 million. (G)

Before undertaking a commitment of such magnitude there has to be a reasonable chance of success. Some types of work are very difficult to break into. For prisons, for example, there are already four firms exclusively engaged in prisons and they clearly have the advantage over newcomers. There is strong competition on education projects because the projects are simpler and smaller so that more SPV groups can qualify. For large projects there may be problems in obtaining sufficient companies to compete.

A contractor is not necessarily involved in the SPV but the awarding authorities prefer that there is a contractor in the group because it creates a unity of interest and purpose in the SPV. (G)

The idea of running the project over a long period of time – say 30 years – whether it be PFI or any other type of procurement where there is commitment of the constructor to operate the works, brings a totally different attitude to the project. One example quoted is that of a pump manufacturer. The normal life of a pump may be 15 years. The SPV asked the pump manufacturer "what do you have to do to make the pump last 35 years instead of 15?" The answer was "I just have to change the nylon bushes to brass bushes. That costs £4 per pump and it will last for 30 years. But nobody wishes to pay that extra bit so I cant sell them." However, the SPV would be interested. (G)

When it comes to organizing the project it is politic to get the design team and the specialist contractors involved at a very early stage but rarely is there any contractual commitment before the point of financial closure. After that point the financing of the construction work is the same as in any construction project – on the basis of monthly payments.

The discussions made it clear that the person who described the process thinks that PFI delivers a more suitable product than the traditional approach. Considering hospital building specifically, the gestation period for a traditional hospital is about seven years and with PFI you would get it in four and a half years. Under the PFI system you have to convince the users of the hospital that it is going to be of the standard they want. The only reservation is that in say 15 years time when the hospital needs renovation the SPV is in a monopoly position to lay down its price. Hopefully the industry will not take advantage of this situation.

One of the problems of arrangements which span more than 30 years is the inevitable risk. The calculations necessary involve such basic uncertainties as the rate of inflation and trends in interest rates. It is the same whether the concession is kept or sold because any buyer will look carefully at the income and expenditure streams to determine the price that should be paid. People are attracted by the enormous turnover of these jobs over their life. The margins are often quite small so figures do not have to be very wrong to wipe out any profit. (H)

Once the PFI project is set up, the way procurement takes place is still open. Against the background of the project already agreed, it could be by design and build or general contracting or any other procurement method. The point has been made that design and build for a project where the client is also the operator, responsible for 30 years, will be more expensive to tender for and to design than one where there is no long-term commitment. This is because the client will have more questions to ask and want more input into the process. This is another variable in costs of transactions. (E)

In undertaking any project in which there is a concessionaire as well as other projects, it is possible to use partnering. Partnering is defined in section 2.2.2 as "the idea of basing contracts on concepts of trust and co-operation instead of price competition". Competition of some sort is usually involved in setting up partnering in the first instance. Partnering operates in various degrees and at various levels. In many projects the main sub-contractors may be brought in at an early stage and participate in the marketing of the whole project. No price can be fixed at that point because no details are available but the sub-contractor's commitment is required for pre-qualification purposes. Over-pricing is not in the sub-contractor's long-term interests because of the need for the relationship to continue for the purposes of securing future work. There has to be trust.

For many companies, partnering has emerged as a means of getting rid of some of the disadvantages of traditional methods of dealing with two layers in the supply chain. A specialist contractor who has an agreement with a supplier to be chosen at every opportunity, they can develop an understanding on reasonable prices, for example, by agreeing prices for a year ahead, and use electronic communications, both of which save a lot in transaction costs. This replaces the need to go to tender for each project to several suppliers and endless arguments with each party once the selection is made trying to grab a bit more of the margin which does nothing to reduce the overall costs of construction. (C)

At the level of supply of components very little partnering is taking place. The idea as propounded by the Latham and Egan reports has, however, been adopted by one or two firms and they are being proactive in forging partnerships. It involves a great change in relationships and trust. Partnership may be more viable if there is added value, in the form of advice based on expertise, which a person lower down the supply chain can add to a product. (C)

One of the public sector clients who had previously worked for a large retail organization and for an oil company said that he always preferred the partnering route. The advantages are that you can work with the architect and the contractor over a longer period of time and you develop the design with both of them. On one of his projects, two large contractors had worked with the design team for eight or nine months before starting on site. At the end of this period the contractors have to sign up to buildability.

At least one contractor is sceptical about the benefits of partnering as a means of procurement. He fears that contractors may see it as a means of increasing their profit margins. If the contractor is sharing gains with the client it should be on a 50:50 basis. He acknowledges, however, that it works on a long term basis where there are real benefits in setting up partnering arrangements at various levels in the supply chain. (D)

The need for both parties in any partnering arrangement to benefit and the need for long-term arrangements in place over a succession of contracts are views common to most commentators.

If there is no long-term agreement, partnering is really a particular form of negotiation. In negotiated contracts, great care is often taken to choose sub-contractors and to make fair arrangements with them. The costs of procurement for negotiated contracts are about double the cost of putting in a tender. This is because there is a dialogue over a longer period of time. Great care is taken to get the contract right. Once the contract is signed, the management time on negotiated contracts may be less because of the preparation they have done, but even so the dialogue continues. It is still cheaper to get work by negotiation because in tendering the success rate is, say, one in four. (J)

A framework agreement is a specific arrangement whereby partnership continues for several projects but usually with a choice of a number of pre-selected contractors and other parties to the process. ProCure21 for smaller health service projects for which PFI is not being used, is an example of a specific type of framework agreement.

ProCure21 is a scheme under which contractors compete to be in a team from which selection will be made for each hospital project. One pilot project started with 51 competitors including nearly all the main contractors. At the end of Stage 1 the field was brought down to twelve. At Stage 2 there were interviews as well as two day visits to sites. There were varying levels of expenditure by the contenders. Some of the twelve used executive jets to fly the selectors and their teams around, others used coaches. After this stage the number reduced to five preferred bidders. This is the group from which tenders for specific projects will be accepted.

Stage 3 is the selection process for a specific project. It consists of an open day with presentations and discussions. At the end of this process the selected group gets a Pre-start Agreement. The selected group says what it expects to spend, though they are not held to this. They then develop the scheme. The client pays for this once the pre-start agreement is in place. If the scheme proposed is not acceptable, the client can go to one of the other five who may use the work already done by the initial group. They may also help the Trust through the approval process. A Guaranteed Maximum Price has to be developed. The NHS initially decided how much schemes were likely to cost but the out-turns were more because cost models were out of date.

The group of five preferred bidders will do contracts of somewhat under £300 million; six schemes each. Most are £2-15 million. One is £55 million. They expect contracts of £30-35 million to be awarded to them, say £20-25 million annual turnover. The first ones started early in 2003. (K)

Each bidding group consists of a number of specialists, one group, for example, consisted of six specialists: main contractor, specialist M&E contractor, medical experts, quantity surveyors, medical suppliers and facilities managers. Costs are substantial in terms of manpower and work may be spread over 1½ years. One successful group was relying on their approach as a well-integrated team to keep costs down. (K)

In a situation where there is selection of, say, three to twelve firms to be on a panel of contractors for a period of say, three to five years some firms are concerned that they are in danger of becoming too dependent on one client for their work and that, once the period of the agreement comes to an end, they could suffer a large reduction in turnover if they are not successful in the next round. Other events could cause problems, for example, the client reducing a planned building programme or even going out of business. In response to these fears some firms have limited their participation in such schemes to a specified percentage of their turnover.

In some of these arrangements each member of the group of selected contractors is told to collaborate with the others in the development of better ways of working.

3.3.4 Selection methods

Once the general scheme for the procurement is determined, then the method of selection for the parties actually undertaking the project must be decided. Some of these have already been described in dealing with the workings of PFI and ProCure21 because it is part of the total process for these procurement methods.

The various selection methods are defined in section 2.2.2. First the choice of the professional team will be discussed followed by that for the contracting team.

It seems that the choice of architect and/or engineer, quantity surveyors and other professionals may be a more relaxed business than for the contracting part of the industry.

One architect was involved with a design competition for a project in Germany. German competitions are much simpler than the British ones but for this project there were 220 initial expressions of interest which were reduced to a shortlist of 15. The client asked for drawings and photographs of three previous competitions – winning projects. "There is more information in a drawing than in 100 pages of text." In addition, he asked for the project strategy and that is all that is required for the first stage. The architect concerned said he would never participate in any design competition where the client asked to have the design up-front. (F)

A quantity surveyor, acting as a consultant on a number of matters concerned with projects, said that a major part of his work comes from other parts of his organization which pass on existing clients to him for specialist help. Otherwise they decide what business they want to obtain and hunt for it. They do not do much advertising. For the firm as a whole he estimated that about 40% of their work is repeat business or obtained by negotiation. Of the tendered work, they are successful in one in three or one in five of tendered projects. However, for the purpose of the survey it is necessary to know this as a proportion of turnover because of the wide variation in the size of projects. (H)

Another quantity surveying firm said that, because they had been forbidden by the RICS to advertise in the past, they still have difficulty in getting staff to actively market the firm. They do market themselves by meeting people and talking. In any case they get a lot of their work by repeat business but some of this would have to be tendered. They also put in some submissions against competition. There are normally two or three other firms making submissions but sometimes only one. For framework projects clients tend to ask three or four quantity surveyors to submit a bid but, in the case of Railtrack, it is four to six and for London Underground six. (A)

Quantity surveyors said that the actual commercial cost of negotiating scope and fees was trivial and so it could be ignored. An architect, in choosing, say, a structural engineer will set up a competition by writing to three or four engineers with whom he has worked and asking them to give him a fee bid on the specification, in relation to the ACE recommended fee conditions. This process may take half a day of the architect's time.

No matter what the selection method, one health service client was not happy about the contribution made by architects on his projects. This is because, even though they had experience of health projects, they cannot fully understand that business, or indeed any clients' business. They are not living with it all the time and cannot devote all their time to one particular type of work because they would not have a continuing work load (I).

The choice of the main contractors is generally more formalized in the traditional process. Until recently the lowest tender was the normal way to select a contractor, for example, in the case of NHS work. Some of the persons interviewed were very critical of the waste in traditional contracting. One contractor listed the following wasteful practices:

1. Clients calling for tenders to test the market, with no intention of placing an order.
2. Lack of proper documentation, which then necessitates assumptions by the contractor, qualifications to tenders and subsequent negotiations and revisions.
3. Inadequate time for the preparation of tenders.
4. Unwillingness of client to commit to a project including disclaimers by client and consultants about information given leading to contractors guessing, qualifying, adding risk etc.
5. Excessive tender lists. They are not interested in design and build for more than four but the Home Office has asked for six.
6. Inclusion in tender lists of incompetent contractors.
7. Naming or specifying sub-contractors (especially M&E), though rarely will this have been arranged or discussed with the sub-contractors themselves. If they are not interested it causes confusion.

8. Inability of consultants to meet their own (or their client's) deadline for sending out invitations to tender leading to lack of information and reduction in effective time to tender while documents are awaited.

9. Failure of client to publish at an early stage the numbers of tenderers invited. Names would also be helpful, see 5 and 6 above. All this information can be obtained but too late to assess whether the firm should submit their own bids.

All these lead to additional costs, especially transaction costs. There was therefore considerable dissatisfaction with the more traditional approaches. (D)

There are also complaints about newer forms of procurement. An instance of this was that there were too many hurdles to getting on to the lists in framework agreements. It was said that ProCure21 had a 200-page document on supply chain management. (D)

A supplier of electrical products mainly to M&E contractors calculates that only one in five quotes turns into an order. They do not quote if the quote is written tightly round a competitor's product or if it is for a named client who normally uses competitor's products. (C)

Some of the NHS clients welcome discarding selective tendering and the NHS rule that they had to accept the lowest bidder. They had to have five or six tenderers. In one case this was disastrous. The successful contractor was just three weeks into the £15 million contract when he went out of business. It caused massive problems: delay, costs, counter-claims, etc. (J)

In another instance they had to accept the lowest bidder from a joint venture. The bid was very low. The contractors had a bad management team. In a 2½ year contract the senior project manager changed three times and the team below him changed too. Each time the team changed knowledge was lost and each time they tried to catch up. They didn't admit all the problems they were having and all the arguments with the client were left to the end so a lot of effort is now being put into dealing with claims. The end product building was all right but it was finished late and this delayed other projects on the same site. With hindsight the quantity surveyor should have been asked whether it was a valid tender or not. This contract illustrated the lowest bidding contractor problem. He is always starting off from a low figure. There is a market level. If a bid is just below the market level, that is OK but if it is far below then it is a bad tender. (I)

The reasons for joint venture bids are several:

1. To share the risk for large projects.
2. To obtain special expertise.
3. To bring in a company to your bid who might otherwise bid in competition.
4. In overseas work, to have a local partner. (E)

In one hospital project, which was a two-stage process to develop a partnering agreement, the client started with six joint venture bids. Each was given a package of information and asked to put together a guide price of roughly how much they felt it would cost to build, who was in their team, how they would go about the project and how long it would take. It was basically an open submission about the project. Then each submission was discussed with the company and two were selected. From these two, one was finally chosen. That is the point at which detailed work starts with the design team and the process lasts several months. The expenses of the selected contractor are met up to the contractual stage but with no obligation on the client to proceed. Assuming they proceed, there are detailed negotiations on the final contract taking in all the overheads and profits to arrive at a price and a contract. (I)

For much smaller projects on a traditional procurement route the same client would like to have six tenderers and their standing instructions require a minimum of three. It is sometimes difficult to get the minimum number so that some companies are persuaded to tender but do not put in a serious bid. (I)

The fee to be paid to the architect would be marginally higher for a partnering project because he or she would have to spend time talking to the contractor. However with partnering there would hopefully be less work at a later stage, so the architect spends more time earlier in the process but less later on.

It is sometimes difficult to tell at the early stages whether a project will develop into partnering. The border lines are blurred. One specialist constructing organization, the oldest and most respected in its field, says that about 60% of tendering enquiries are either two-stage or negotiated. The other 40% are generally reduced competition, i.e. no more than three tenderers. The process may start with a telephone call inviting them for an exploratory chat. The so-called negotiation that follows is not just negotiation; it is the evolution of the design. In describing the process for a particular project the contact told us: "As the design team churns out the drawings, our role is to look at the budget to see if the proposal will take it over but if they really like it then we have to try and find savings elsewhere, so there is always a balance. There are things we would like, if we can afford them, but there are certain things we <u>must</u> have. On that basis it is always keeping tabs and it is a very transparent process. We put it all on spreadsheets; it is marked up, labour and materials, all mark-ups and everything else. So every time a set of drawings is changed, the quantities are amended and items added. The bottom line reflects it all. It is e-mailed back to the client who goes through it and whatever happens we get to a point from when we started the job, where the project is in line with what they wanted and, in this case, came out exactly at the cost estimated". (B)

In partnering all sorts of apparently minor problems can be avoided. For example, if specialist bricklayers are working an area of walling and find that other trade contractors are working in close proximity, the bricklayers may have to build overhand so it reduces productivity. If it had been a partnering job that would have been foreseen and avoided. (B)

3.3.5 Commercial costs at various stages in the process

The consultants

Almost all professionals do some form of marketing, but how much and by what means varies greatly. A small architectural practice did virtually none, just relying on word of mouth (F). A large professional QS practice stated that it does think about marketing though most of it is achieved simply by meeting people and talking to potential clients. They do not have a marketing strategy. Marketing may represent 1% of their costs but it is very difficult to put a time to it because any dinner or lunch with someone is a form of marketing. The percentage of time spent by partners and other senior executives may be, say, 10%. There is still a hangover of attitudes from the time when the RICS forbade its members from advertising and that means that it is not a traditional activity of the firm. (A)

The question was asked "when does marketing stop and putting in a submission begin?" The latter initially may be a case of sending information about the firm but then it develops into discussions about a project. The cost of putting in submissions has been estimated at 2.5% of payroll for this firm. (A)

One of the problems for professional firms obtaining work is that their commission may start with one part of the total service available and then proceed to another and another. That means the firm may have to go through the negotiations on service and fees say, three times. That is very wasteful. (A)

Some firms do not distinguish between marketing, the preparation of submissions and the follow-up on the submissions. One partner of a large firm said that he spends 20% of his time on these activities. That is the amount of time he has to spend to bring in the income to feed four to ten people with work. He commented that recently he had not been doing that and it is evident that there is a danger of shortfalls appearing in the future. There is a constant conflict in time between getting the work and doing the work. (H)

When it comes to Stage 3 of the commercial process, namely monitoring performance, the objective of the research is to find out how that varies with the type of procurement method used. However the point was made very strongly that for professional firms, it is the way the main project is procured which determines whether the costs of monitoring are greater or less than average, not the method of procurement of the services from the professional firms. An example was quoted of a project which was going to be a single stage, simple tender but the client decided to turn it into a target cost, construction management project. The role of the professionals and their fees increased substantially. For most projects monitoring of the firm's own professional staff is a normal activity no matter how the project has been procured. As soon as the method of procurement of the project changes then monitoring costs change dramatically.

Under framework agreements with Key Performance Indicators (KPIs) there has to be continuous monitoring. The client may think that savings can be brought about by using KPIs because it avoids having to keep going out to tender to test the market. In fact, the hassle for the professionals of administering the KPIs means that they may be costing the client more, rather than less money. One problem is that the client thinks that because performance is being monitored, the work on site is well done. In fact the contractor may not be efficient. (A)

At Stage 4 when the project is complete there may be costs of dispute resolution. There are two main types. One is where there is disagreement as to how much money is owed and the other is where the amount is not in dispute but the client, of whatever sort, has not paid In the case where there is a dispute on the amount due, there are many levels of acrimony. It may be that you say to the client "the brief has changed and we would like to review our fee in the light of that change." He may say the brief has not changed and you say all right, let us sit down and decide where we are. That, said the interviewee, is not really a dispute, just good management. The acid test of a dispute is whether the parties are so entrenched that there has to be a third party involved in the settlement. That may be resolved by an independent person, through mediation or conciliation and only in extreme circumstances in court. One quantity surveying practice thought that less than 10% of projects, probably less than 5%, suffered real disputes. They have had situations where they have claimed on their professional identity (PI) insurance. Consultants tend to regard their PI policy as a resource they can tap into. This firm does sometimes do work for contractors, sub-contractors and suppliers. The contractors and suppliers tend to be more commercial than clients in their approach but they do so little work for the former that it is difficult to say. (H)

Another large firm estimates that it is spending about ½% of costs on debt recovery, i.e. two people full-time plus time put in by individual partners. In a period in which the economy is in a good shape it would be less than in bad times. Cash flow is vital to consultancies: 20% of turnover is outstanding more than three months. The worst ever client in this respect was Government.

Contractors, trade contractors and suppliers

Marketing, for most contractors and suppliers, is based on personal contacts. The companies interviewed were all respected in their fields, with excellent reputations, so that they are often an obvious choice at least for a preliminary enquiry. One specialist contractor said that every individual is expected to market the company. One director will go to various trade associations etc. However, there is no sales team on the road, no marketing director, they do not wine and dine or have golf days. As one of the first companies in the field they get their work purely by invitation. (B)

A large contracting group stressed the need for building relationships and this is done at various levels. However, another member of their team thought the informal marketing was not very important. This is particularly so with public sector clients. They will not accept invitations to dinners or other activities for fear of being accused of colluding with the contractor. (E)

The owner/director of one supplier of complex components is working hard on getting partnering arrangements accepted as a major procurement method across the industry as a whole He believes that that activity has brought a great advantage to his company in obtaining business. Another form of marketing is to be helpful to consultants about the appropriate technical specifications. They then put that in their overall design. The consultant has the responsibility for it, but the costs of doing the calculations have been borne by the supplier. It may be a marketing cost but, because they may practically have the job, it is, in a way, the cost of doing the job. However, marketing costs need not be chronological so it can still be categorized as marketing. (C)

One of the main contractors interviewed keeps good records on the costs of procurement. He regards marketing costs as 16% of the cost of procurement, estimated on the basis of personnel costs: salaries, bonuses, cars etc. They do not include normal office costs. They have three people plus the equivalent of two secretaries on marketing, plus various directors' involvement. The turnover of this company is about £80m and it is estimated that procurement costs are about £1.1m or 1.4% split as follows:

	%
Marketing	16
Estimating	38
Design and Build Dept	21
Planning	9
External Costs	9
Legal Internal	4
Legal External	4
	100

The design and build department costs are high because of the design input. On average they win 1:3 or 1:4 of their design and build projects. For ordinary tendered projects it may be 1:8. Negotiated projects, although fairly certain, require a substantial amount of estimating and discussion. Not all negotiated projects lead to contracts, largely because the client does not proceed with the project. (D and J)

Costs of procuring suppliers

Just as clients have costs of obtaining tenders for their work, so contractors must obtain sub-contractors and suppliers and so on down the supply chain.

The contractor that estimated its procurement costs of getting work as £1.1m also estimated the costs of procuring suppliers and sub-contractors as nearly £1m. Their normal practice is to use competitive bidding for both procurement types. On average this is about four or five tenderers, ranging from two to eight. One of the reasons why it may be a high number is that for some materials, such as structural steel, the manufacturers have to keep their production going on a fairly continuous basis. Any one manufacturer who is short of orders will bid a low price to keep the factory operating. If the contractor goes out to say eight producers, one of these may want the work badly and the bid will be correspondingly low. At the other extreme are, say, mechanical and electrical engineers who may have a substantial input. They need to have a lot of discussion before they bid and it is vital that they are on the same wavelength as the main contractor. (J)

One contractor observed that the cheapest tender to carry out, in terms of procurement costs, is a "construct only" tender where there is a bill of quantities provided. The next is "construct only" where they have to produce their own bill of quantities. Much more expensive is a design and construct tender, not only in terms of their manpower, but also because they have to commission a designer. Even more expensive is the design and build contract for a concession company or for a PFI job because then they must take into account whole life costs and their requirements. (E)

3.3.6 Other observations

There were a number of interviewees who were musing about the real, long-term effects of new ways of working. Some of these are reported below.

One professional observed that some of the modern procurement methods will actually increase the cost of building and much of the new procedures will be unsuccessful. Traditionally some contractors went bankrupt very quickly because they rolled a loss over from one job to the cost of the next one. But over the last couple of decades building costs have not gone up. Efficiency and other factors have kept a little bit of profit in for contractors but there is not enough. The giant contractors of this world are going to have problems. The contracting companies they have bought were never supposed to be profit-making. They existed purely to give their directors some income and, if there was anything left over, they shared it between them, but they were private companies. They were not profit-making organizations. Companies cannot spend two years tendering a PFI, with millions being wasted on tendering procedures unless it is recovered. (A)

There has been a major change in the way quantity surveyors are used and their contractual arrangements. At one time the rules were laid down by the institution. If you appointed a QS you got the standard services and the price was fixed. One problem with this was that the harder you worked to get the price of the project down, the lower your fee, since it was based on the cost of the building. Now you start with a conversation with the client and then work out your price. If that is not successful, you may have to omit some services and then the QS becomes very commercial. Within one organization there may be project managers, quantity surveyors and building surveyors. Is the advice as to the best method of procurement influenced by the interests of the other parties? (H)

At the Latham and Egan level, there is a real climate change and there are a lot of people interested but it is not penetrating down the industry. The advent and use of frameworks and partnering, in themselves, indicate that people are unhappy with the status quo and are trying to find an alternative that will deliver. But I think there are many clients working on the basis that "it won't happen to me". They may even do some mental risk analysis that says "OK, I know I get it at this price instead of that price and that one project goes wrong every now and again. But actually, as long as only one in ten goes wrong, it doesn't matter because I am getting the other ones low enough to make up for that one, whereas the new methods might cost more." They are playing the odds. One of the best things that has come out of the industry in the last few years is legislation which deals with malpractices such as "pay-when-paid" and withholding money. (H)

Thinking about overall change in the industry, one comment was that partnering, alliances and frameworks are almost trying to get back some of the continuity of relationships which were there in the 1950s and 1960s when contractors retained their own labour force. If there is not enough continuity for contractors to keep their own labour force, are we kidding ourselves in thinking that there is sufficient continuity to have meaningful alliances and partnering agreements? (H)

One contractor is concerned that the various reports produced with Government: Latham, Egan and now ProCure21, may be doing a disservice in creating expectations beyond an achievable level: "They are trying to increase value for money and reduce cost. However the physical materials and components have a cost which is largely fixed. Any improvement must come from the process of erecting them and putting them together. One needs also to think about the costs before a building is decided, e.g. planning applications and land acquisition." (J)

Why should partnering work? The client wants the best price, the best design. This may be possible, but if it is virtually imposed on the contractor, it is dictatorship not partnership. How can you share risk with the client? An architect who makes a mistake in the design, and a window is in the wrong place, cannot afford the cost of putting it right if there is no margin in the fee. The architect gets paid very little, and only for labour. Partnering may not be cheaper. At a lecture, a BAA person said that in spite of all the problems of collapsed tunnels, etc., the project was completed on time and to budget. How can this be – unless the budget and allocated duration were ridiculously high in the first place? The same applies to negotiation. The client wants a good deal and the contractor wants a profit. The contractor likes negotiation. It is simpler and easier but not necessarily cheaper. Just better for me! (D)

What does a sub-contractor gain from having preferred status? Perhaps too many eggs in one basket if all the work is for one contractor. For example, one company has a sub-contractor currently requesting work. He has done a lot of work for them and was so busy he did not work for others. Now that particular type of work has dried up, he is not getting work elsewhere. He does not have the right track record with other contractors and they may regard him as a protégé of the one contractor. Too much reliance on one client is dangerous, as circumstances change. The maximum desirable workload from one client is 20-30%. (J)

ProCure21 is looking for benefits from partnering also. One company spent £¼m trying to get on to a pilot project in the North West but did not succeed. Now they are submitting again, but it is difficult to see where savings are coming from. Why is there not more standardization in the hospital programme? A hospital bed requires some space round it and certain services and this requirement is the same in any part of the country. If they standardized production they could get greater efficiency in design and production. Maybe operating theatres would need more variation but most hospitals have similar needs. (J)

Key performance indicators (KPI) are supposed to lead to continuous improvement. You must: collect data (which is expensive), assemble data and compare with norms or historic data. Then what do you do about it? They use KPI for safety checks. (J)

You can not go on driving costs down by 10% per annum. Supermarkets have got costs down and, especially, time of construction has seen dramatic falls (from 40 to 18 weeks for Tesco). They have done this partly by the method of construction of the external wall. They use a jumbo stud partition with insulation and waterproof sheeting with terracotta tiles. It might be unacceptable for other building types. They do an internal revamp every five years anyway. Their buying power is enormous. It is debatable whether their system is partnership or dictatorship. Tesco set a price for certain items – the same for everyone, for example up-front costs and certain overhead items. This simplifies the tendering. In spite of all this, he thinks there are some good things in Egan. (J)

3.4 TIME SHEET AUDITS

In the early stages of the research, an attempt was made to monitor the activities of people involved in the procurement process. This was intended to be a cost-accounting exercise, to find out exactly how many hours each person spent on the activities associated with procurement. The aim was to distribute time sheets, according to a pre-determined format, to people throughout the supply chain. A large sample of data was required, and that would enable the characterization of the typical costs associated with certain types of project, in certain locations. Simply defining the variables turned out to be a huge task, especially with an objective of providing limited options for respondents to choose from, so that data from different companies would be comparable.

3.4.1 Method

The survey was split into parts. The first part was intended only for one of the senior managers of a company, who would provide the name of the firm, annual turnover, typical role (funder, developer, client, main contractor, sub-contractor, trade contractor, supplier or consultant), type of business (e.g. property developer, glazing contractor, architect), sampling period. Then data were required for each individual in the return: name (or ID number), job title and gross annual salary.

In completing the pro-forma shown in Table 3, the aim was to have the respondents choose from limited ranges of options. The options were developed during a series of interviews with practitioners from across the supply chain who shared their views about the kinds of option likely to be needed. This provided an interesting overview of the range and variability of procurement options from the perspective of many different participants in different kinds of company. The options for their responses are shown in Table 4 (the "key").

Table 3: Data entry form for individuals in a main contractor's organization

Name or ID No. [ ] Period of data: [ ]

Nature of the projects during this period (please see key for choices):

		Projects worked on this period					
		A	B	C	D	E	F
Type of client	Nature of client						
	Frequency of client's experience						
	Origin of funding						
Project descriptors	Value (£ thousands)						
	Duration of construction work (weeks)						
	Type of building						
	Size of project						
	Location						
	Extent of main contractor's design responsibility (%)						
Procurement of the project	Basis of relationship between client and main contractor (or CM)						
	Method of procurement						
	Number of bidders						
	Extent of sub-contracting of main contract (%)						

Nature of the transactions with your suppliers and sub-contractors during this period (please see key for choices): We are interested in your supplier and sub-contractor relationships to the extent that they relate to the projects above. If any of this work is not attributable to **specific** projects, please enter "X" for the project code, otherwise use the letter (A, B, C … etc.) relating to the project identified in the table above. Each of the numbers (1-10, and 11-20 on the continuation sheet) relates to a particular transaction with a supplier or sub-contractor. It is likely that each project will involve several such transactions, in which case the same project letter (A, B, C … etc) would appear against several of the transactions numbered below.

		1	2	3	4	5	6	7	8	9	10
Procurement of your suppliers and sub-contractors	Project to which supply relates (A, B, C etc)										
	Relationship with sub-contractor										
	Type of transaction										
	Value of this transaction										
	Number of bidders										
	Extent of design responsibility (%)										

No. of hours spent on each stage in relation to your customer/client:

Hours:	Non-project work	Project hours for this period					
		A	B	C	D	E	F
Stage 1: Marketing costs							
Stage 2: Tendering/negotiations, price & scope, specific project							
Stage 3: Participation and advice on design							
Stage 4: Contract administration and adjudication							
Stage 5: Finalizing payments, settling claims							
Stage 6: Arbitration or litigation							

No. of hours spent on each stage, in relation to your suppliers/sub-contractors:

Hours:	Non-project work	Project hours for this period									
		1	2	3	4	5	6	7	8	9	10
Stage 1: Search information costs											
Stage 2: Negotiations, fees & scope, specific proj.											
Stage 3: Participation and advice on design											
Stage 4: Construction, cont admin & adjudication											
Stage 5: Finalizing payments, settling claims											
Stage 6: Arbitration or litigation											

Table 4: Key for data entry form

Type of client for whole project

Nature of client	Frequency of building experience	Origin of funding
De-Developer En-End-user Or-Other	On-One-off Oc-Occasional Fr-Frequent	Pr-Private sector Pu-Public sector PF-PFI or PPP

Project descriptors for whole project

Value	Duration	Building type	Size	Location	Main contractor's design
£000	Please enter approximate number of weeks for the overall construction period, estimated if necessary.	T1 Housing T2 Roads, harbours & railways T3 Water & sewerage T4 Industrial T5 Schools & Universities T6 Health T7 Offices T8 Shops T9 Entertainment T10 Refurbishment, repair or maintenance T11 Other	T1 No of units T2 Length, or other unit T3 Sm, med, large T4 Square metres T5 Square metres T6 Square metres T7 Square metres T8 Square metres T9 Square metres T10 Sm, med, large T11 Sm, med, large	L1 Scotland L2 North L3 Yorks & Humberside L4 North West L5 East Midlands L6 West Midlands L7 Wales L8 East Anglia L9 Beds, Essex & Herts L10 Berks, Bucks, Hants & Oxon L11 Kent, Surrey & Sussex L12 South West L13 London L14 Continental Europe L15 Rest of the world	In pure design and build projects, the main contractor's design responsibility will be 100% as the liability extends over the whole project. In pure general contracting, it will be zero. In practice, the number may be somewhere between.

Procurement of the whole project

Basis of relationship between client and main contractor (unless CM)	Method of procurement	Number of bidders
Pf-Preferred or framework Pa-Partnering Ne-Negotiation 2s-Two-stage tendering 1s-Single-stage tendering Oc-Open competition CM-Construction Management	GC-general contracting (construct only) ND-novated design-build PD-pure design-build CM-construction management MC-management contracting Or-other	How many contractors were invited to bid for the whole project? Enter PT if the project is at a pre-tender stage

Procurement of your suppliers' or sub-contractors' contributions

Relationship with supplier of sub-contractor	Type of transaction	Extent of design responsibility	Number of tenderers
Pf-Preferred or framework Pa-Partnering Ne-Negotiation 2s-Two-stage tendering 1s-Single-stage tendering Oc-Open competition	Ad-Advice De-Design SM-Supply of materials SC-Supply of components In-Installation La-Labour	For what proportion of their work on this project have they design responsibility?	Number with whom the supplier or sub-contractor is/was competing Enter PT if the project is at a pre-tender stage

3.4.2 Findings from time sheet survey

Although the steering group was particularly helpful in enabling the development of a systematic approach to the collection of time sheet data, and even though a lot of collective effort was spent in developing this approach, in the event, almost no one was able to provide the data requested. One of the returns had a covering letter which included this text:

> *Most* [of the respondents] *were concerned that the sampling on the form did not give a very useful view of an overall picture. Some also found the forms awkward to fill in because the information required was not aligned with what they normally thought about.*

Given these reservations, it is remarkable that any responses were received at all from this company! With reluctance, this approach to data collection was abandoned. While it provided no data that could be analysed, it helped to drive many discussions with industrial partners and these provided tremendously useful insights into the nature of the problems being researched, and these insights informed the rest of the research. This was also a valuable insight into the way that people in industry react to random sampling.

3.5 STRUCTURED INTERVIEWS

A series of structured interviews was carried out with various firms from parts of the supply chain, to find out what they thought were the most important aspects and problems associated with getting work. Fifteen structured interviews were carried out, and in the summary of responses use the following key to identify who said what:

- CL1: Grocery chain
- CL2: DIY chain
- CL3: Health trust
- CO1: Civil and structural engineer
- CO2: QS Property advisor
- CO3: M&E consultant
- CO4: Architect
- CO5: Architect
- MC1: Building contractor
- MC2: Building contractor
- MC3: Building contractor
- MC4: Building and civil engineering contractor
- MC5: Building contractor
- SC1: Specialist trade contractor
- SC2: M&E contractor

3.5.1 Trust in the industry

Do you find people in the industry generally trustworthy, or not? Is this any different to your views about people, generally? Is there something specific about those who populate the construction sector that predisposes them to behaving in any special way?

- CL1: Ten years ago people in the construction industry were definitely less trustworthy than people in society generally. But now, things are better, helped by collaborative working, which has only found its way into a small part of the industry. Construction industry people are as trustworthy as people in society generally.
- CL2: Most people in the industry can be trusted, as in the population, generally. You need to engage with them in the right way. I don't like open book accounting; it is too easy to manipulate and takes time to look through people's books. I prefer to take people on trust in terms of them delivering what they have promised, and being paid.
- CL3: Our experience is varied. The worst was a joint venture project that was a failure because the contractors were not trustworthy through the building process. They did not programme properly, and did not resolve problems, leaving them all to a difficult claim at the end.
- CO1: Generally quite neutral, although a wide range of experiences, including architects who are suspicious about losing their clients to other consultants. When it comes to claims people have no interest in partnering.
- CO3: I think we have a disproportionate share of rogues in the construction industry. In competing with specialist sub-contractors offering D&B, they may promise more than they can deliver and then ask us to help them deliver what they said they could do without us. Although it is changing, the industry is still very adversarial. In negotiated arrangements, there is a danger of prices creeping up.
- CO4: Most people in the construction industry can be trusted.
- MC2: With partnering, there is trust and consistency of performance.
- MC3: We often have to provide a non-collusion certificate to confirm that during the tender stage, we have not communicated with other contractors. I suppose that clients worry that if we know who else is tendering, we will collude. That is a sign of a lack of trust. It is treated as a big secret even though people in the industry know it is not a secret: it is just a farce.
- MC4: I think the people are the same. Commercial pressures may persuade people to behave less honestly. One fraud we have long come across involved doctoring invoices for variation orders. Partnering has got to be based on trust, because what we say during tendering can only be proved after the event.

- SC1: We are selective in who we tender for. We would check a new sub-contractor or customer with Dun and Bradstreet. Most of our clients are well known to us and we just don't bother; perhaps we should be, given the state they are in. If it was completely an unknown client, with an unknown or dodgy main contractor then we wouldn't bother tendering anyway.

3.5.2 Preferred procurement - Marketing

How do you deal with marketing (or if client, discovering potential contractors)?

- CL1: We have pared down our list of external consultants and contractors. It is not completely closed because new people have to be introduced. The biggest problem facing the construction industry has been a lack of continuity.
- CL2: We have a shortlist of contractors with whom we have worked in the past, on a quasi-partnering basis, and we seek to market-test them every two years.
- CO1: We market to property developers, architectural practices and main contractors, and to a lesser extent, Local Authorities. Our marketing is networking, i.e. being aware of the marketplace and pitching for selected projects. We have a lot of repeat work and some partnering arrangements.
- CO3: Cold calling never generated a lot of business. It all comes down to getting your first job with someone and performing well. We don't spend a great deal on marketing. We try to keep in touch with people we know and as they move, we keep in touch with them. We do the same sort of marketing that other people do, entertaining clients and so on.
- CO4: There is a sales and marketing partner and a new business executive whose sole duty is business development and marketing. We have a business plan for targeting sectors and amounts. We carry out systematic market research, direct mailing and public relations. We are in pre-qualified lists such as "Link up" and the RIBA client advisory service. We respond to OJEU notices. Most important, is networking, being part of the business community.
- CO5: We spend very little on getting work, because we have just "tumbled" that it is a total waste of time. We get work by word of mouth, which is quite sufficient to keep us occupied. We offer each client we take on £5,000 of pro bono work, which is a marketing tactic.
- MC1: Work is obtained mainly from previous relationships. They are not short of work and are well known to local authorities, health service and housing associations. They do target specific schemes.
- MC2: In our regional office we have a marketing department that also deals with training. Our local marketing staff have to find out the basis of jobs before we decide to tender them, e.g. contract amendments, qualifications, warranties. We try to develop long-term relationships with key clients.

- MC3: We have a marketing department employing a full-time team of people. In addition, various directors are involved in long-term aspects of marketing.
- SC1: Most of our clients are well known to us. Our work generally comes through our relationships with six main contractors. We do not have a marketing department, as such, but we are well known in the business.
- SC2: We have a marketing department, which produces presentation documents and a business development manager who administers all the leads received. The leads come from everyone in the company and their families and friends, including apprentices and their parents and even grannies. They give £100 prize for any lead which results in work.

3.5.3 Preferred procurement - Tendering

When putting a contract together for construction work, which is more typical, tendering or negotiation? Why? Does this differ when you are buying rather than selling (if not a client)?

- CL1: More typical is negotiation and collaboration, contractor-led. Competitive tendering gives you the wrong contractor at the wrong price; it begins with conflict and this infests the whole process, diverting the team's attention to animosity, confrontation and division. Partnering gives the team a much better insight into what they are doing for the client.

3.5.4 Preferred procurement – Tendering - No of bidders buying

What is the best number of bidders when buying construction work? Have you specific experiences to relate about when you have been close to the ideal and just how far away from your ideal have you been driven, and by what?

- CL1: When negotiating, it is one. My preferred number of bidders when I have to tender is four, but my company's preference is 6-10. The furthest that I have been driven from my ideal is to six.
- CL2: In an open tender we would get probably six bids.
- CL3: We have to go to at least three as per our standing instructions, and I like to go to six, for the range, but it is difficult to find more than three or four. Market forces affect the number of people who will bid for our work. We have had as few as two. We sometimes get into the ludicrous position of begging contractors to bid simply because we need a number of bids, otherwise we have to make a case for having less.
- CO1: I prefer to negotiate.
- CO3: As a consultant services engineer we are often asked by a mechanical and electrical sub-contractor to provide an outline design in response to a

performance specification. We will charge for this. If four or five such sub-contractors each do this for four main DB contractors, the services are being designed, for a fee, approximately twenty times.

- CO4: We do it rarely and we would never generally go to more than three bidders for professional services because generally it is very difficult to draw up a list of more that three people who would have the proper expertise to do the job, so you are restricted by the size of the market.
- CO5: When procuring engineering consultants, I invite three or four engineers to submit a fee bid.
- MC3: Generally, the average is four to five tenderers, but it may go down to two or three. Manufacturers with continuous processes, such as structural steel, need to maintain a steady flow of work. If their order book is not full, they will drop their prices dramatically. In order to find these, we will invite bids from as many as eight of this kind of supplier.
- MC4: I think the ideal would be just one. It has to be based on the right kind of relationship and the right control. But we need that relationship where people are honest with each other. In reality it doesn't happen very often. In negotiation, a contractor whose overheads are assured will be much more likely to be working with the client, rather than against him.
- SC2: In sub-contracting and purchasing about one third of all materials bought are partnered. They try to bring in sub-contractors early and negotiate with them. In tendering, they like to have no more than three bidders. They would prefer to work on a one-to-one basis but this may not get a competitive price.

3.5.5 Preferred procurement – Tendering - No of bidders selling

What is the best number of bidders when selling construction work? Have you specific experiences to relate about when you have been close to the ideal and just how far away from your ideal have you been driven, and by what?

- CL2: When negotiating, it is one. The furthest we have been driven from my ideal is to six.
- CO1: Unlike contractors, we rarely know how many people we are bidding against, although we know who is likely. Six to eight is the number preferred on a large project, four to six for a medium. It is ironic that there are fewer bidders on smaller projects because they are cheaper to quote. These numbers are normal but we regularly find it is double. In advising the client on how many contractors to invite, we want confidence that a bidder is actually going to be able to deliver what we have designed. We also have an interest in ensuring the bidders are not wasting their time; e.g., there have been occasions that I have helped to bring a client's list down from 12-15 to 8-10 bidders.

- CO3: The best number of bidders when selling is one. The largest number we have had is probably four. A lot of our work is in competition, but M&E contractors tend to use only one or two consultants all the time so there is not a lot of competition, and it doesn't tend to be on price. We probably lose more work because we can't do it rather than because of fee bidding. We would not dismiss a 1 in 10 chance but we would have to think about that quite seriously.
- CO4: The ideal is one. We have been driven far from that ideal in some cases. Tendering for consultancy services moves pretty much like tendering for construction work and most people do it on the NJCC guidance: the greater the price the greater the number of people they ask to tender. We have generally found that it is reasonably realistic.
- CO5: There was one design competition with a long list of 220 and a short list of 15 which we got into on the basis of one simple document.
- MC1: We will not bid when there are more than six on the list; we prefer fewer.
- MC3: There may be as many as ten requested, usually six, the chances are there would only be three or four serious bids. Some would either send it back or take a cover. But they still incur cost to the industry, even if contractors only look at it, consider it and discover how many are on the list. If they committed never to go to more than four, and checked that all four were interested, the reduced competition would make us more interested.
- SC1: About 60% of our tendering enquiries are either two or negotiated. The other 40% are generally no more than three tenderers.

3.5.6 Preferred procurement – Tendering - Non-competitive pricing

If competition is not the basis for selection, how do you deal with arriving at an agreement on price?

- CL1: Pricing is the normal tendering process of preparing a bill of quantities with a PQS and having the contractor price it. But we know what it should be because of our experience and if their price is too far from this, they know we will just go out to tender.
- CL2: For freeholds, we deal with main contractors. With our fit-outs, we deal directly with trade contractors, on a serial contracting basis, bidding from six, with prices held for the year. With the electrics we have about four contractors on the same standardized schedule of rates. To choose among them, we just spread the workload, according to our requirements (not their choice). By going for volume, we have bought materials directly and reduced the number of middlemen.
- CL3: We negotiate based on the market level as judged by our consultants.

- CO3: A full design specification and site supervision service through to the end of the project usually constitutes 5-8% of the total building services contract works. We do lump sum pricing by working out our costs and comparing them to that level.
- MC1: It is right to get away from cut-throat competition, which is far too costly. Partnering is not the only answer and does not solve all the problems. A possible alternative is cost-plus but the problem is identifying who pays for mistakes.
- MC4: We agree our overheads and profit levels "up front", with the client. Then we know how much money we have to pay for the work to be carried out. There still tends to be some element of competition; we"d invite tenders from four sub-contractors, often at the behest of the client, who still wants these packages to be market-tested. By asking for this, clients are missing the broader benefits that companies can bring, in terms of value engineering skills, strength of their supply chain etc.
- MC5: In one example, we agreed a mark-up for all projects, but this has proved difficult to work to. The result of this is that we now keep separate accounts for each client so we know our real costs, enabling us to decide who to work with in the future. In another example, we work on the basis of a price per square foot, enabling us to focus on the best product.
- SC2: We aim to obtain 50% of our work by negotiation without competition.

3.5.7 Preferred procurement - Monitoring

How do you control progress and ensure that all parties are doing what they promised? Examples?

- CL1: Within our standard documentation we have various schedules: design programme, procurement schedule, construction programme, client approval schedule. All of these are monitored on a monthly basis. This picks up a lot of problems at an early stage. Because most contractors call for information a long time before it is actually needed, often six months before, then even when they are honest about it and give genuine dates by which information is needed, architects are so acclimatized to the habit of it being requested too early that they assume they have plenty of time. These documents help us to avoid this because all of the information in them is available to everyone in the project.
- CL2: The majority of the on-site checking is done by consultants reporting back to the client. In terms of resource for monitoring, it does not vary with the type of procurement. In collaborative working, contractors may see an opportunity to lower the standards anyway; quality is something you need to monitor irrespective of the approach. We have started to use thermographic imaging to test that insulation is properly incorporated into walls and claddings.

Unless you have got somebody on site at the time, monitoring it, there is no way they will pick that up and in most of the traditional routes the inspections are weekly.

- CL3: We employ a management specialist to manage our projects and programme them accordingly and work with the contract administrators to chart progress, check progress and re-forecast where required.
- CO1: We establish a design and information production programme. We use time sheets to get regular feedback on the fee spend. The clients do not have people watching over us because we are the consultants. The only people that ever watch us are project managers. The monitoring is to do with the programme on site and our output related to that programme of works on site and is often driven by the spending of money.
- CO2: Once the main contractor is appointed we do not have much say in the choice of sub-contractors. There are two distinct types of sub-contractor. The first has many direct employees with a good craft base supported by a well-equipped workshop. They generally do good work and are reliable. The second category is where a former contracts manager has set up a company to manage labour-only sub-contractors. They are generally confrontational and difficult to work with.
- CO3: We have a monthly meeting with everybody in the office to monitor our own work.
- CO4: I think the only pressure we can exert on the other members of the design team is through progress meetings and we have deadlines. If the deadlines are not met, we tell the client and that is the point at which some commercial sanction might be brought to bear but that rarely happens. Usually the problem with progress occurs because a job is not correctly resourced. Once you get into the contract stage there would be an information release schedule produced by the contractor.
- MC4: There is less policing than there used to be. There has to be control in terms of time and cost.

3.5.8 Preferred procurement - Dispute resolution

How do you deal with the resolution of disputes and arguments? Examples?

- CL1: Immediately, generally speaking, unless we are not aware that there is a dispute.
- CL2: We very rarely have contractual disputes with contractors.
- CL3: We have a claims situation on a contract where we are offering £4m as the claim figure, which we are happy with. They want £10m and it is going nowhere. So we organized workshops with the two sides together. I chair it

and we go through each issue and try to narrow the gaps. We are trying to avoid arbitration or adjudication, trying to get to a mutually agreeable solution between two parties. At least, if it goes to arbitration, there is a clear list of things that we disagree about. For closing the gap, workshops have been very beneficial. I am hopeful that it will cut down the legal fees in the longer term. Collaboration between the two sides is the only way to close the gap. If the contractor had a strong case, he would probably have gone to adjudication.

- CO1: In some D&B projects, the client may end up with something he didn't quite expect. This leads to a paper chase, and the fit-for-purpose argument often comes up. We do get involved in thorny issues between the client and the contractor. We feed this back into our procedure. Usually a dispute is a claim against our fees. When we have a dispute with the client over payment, we ring up a couple of times and they say the cheque is in the post, so then we will wait for it to arrive. Whereas when we have had disputes with contractors we have gone to court to settle them.
- CO2: We do not have much trouble with disputes. All of our projects are well documented from the start and instructions are clear. This puts us in a very strong position. We have good information management and deal immediately with any problem.
- CO3: We do have problems on jobs but they are generally straightforward, low-level disputes and we have been able to resolve them.
- CO4: Disputes between us and sub-consultants are rare. Disputes between us and clients usually nowadays follow a sequence. It starts with informal negotiation between senior members of the two organizations which then moves on to some sort of mediation involving a third party, which then moves on to adjudication and arbitration. We tend to step up through those processes. The resolution of disputes has to involve the input of our underwriters because ultimately they have to pay. They take a great interest in how disputes are resolved, and there are certain forms of dispute resolution that are not acceptable to them, so this tends to be dealt with at the outset.
- CO5: Arbitration and litigation are very infrequent.
- MC1: We do not have many disputes. There should be even fewer with partnering but this has not been properly tested as yet.
- MC4: In partnering, problems don't arise; they are sorted out during the process because we haven't got a set of documents and a procurement route that sets you off on this adversarial footing. If the process is wrong, then you can still have almost as many disputes on those projects as you do on other projects. They are just dealt with in the usual fashion. I joined the company 3½ years ago. In the first year I think we had about 5 or 6 adjudications, in the last 2½ years we haven't had any. And that basically mirrors the shift in work from the more traditional to partnering and negotiation.

- MC5: Resolution of conflict is much easier and quicker with partnering. The function of the QS is to ensure that the figures are within certain bounds. The objective is to reduce the grey areas which cause additional work.
- SC1: We have the odd spirited debate over the magnitude of payment, but generally speaking we always reach an agreement and get paid. There are problems when a company has gone into liquidation, but that is very rare.
- SC2: There are always some but we keep them to a minimum. One of the problems with partnering is that the client tends to expect that he will get some things free. They get the best of both worlds.

3.5.9 Problems - Marketing

For the marketing stage in the commercial process which procurement method and/or selection method seems to cause the most problems?

- CO5: We don't enter design competitions where they ask for the design upfront.
- CL3: Under competitive tendering, it is a long drawn out process in terms of paper work to arrive at your lowest bidding contractor. At least with partnering, you can get them involved far earlier and far quicker.
- MC4: In forming the relationships, it is much more difficult with partnering, particularly a strategic relationship, because there is a much more long-term commitment to these relationships.

3.5.10 Problems - Agreeing terms

For the agreeing stage in the commercial process which procurement method and/or selection method seems to cause the most problems?

- CL1: Tendering competitively. It involves a very detailed breakdown of the work and a build up of a comprehensive price. There is an inherent problem in the construction industry, sometimes referred to as "director's discount" in which a senior executive can simply look at the bottom line and insist that £1m or some such arbitrary figure is taken out of the tender. The staff just have to make it add up. Again, the decision-makers do not have to live with the consequences of trying to make this work.
- CO4: Some big companies have strict buying processes in place and so they expect the work to be tendered and opportunities for negotiation are limited. Also, repeat business is interesting because often we go through a tendering process which is an empty process because they want us to do the work anyway and they are testing the market to make sure our price is right.

- MC4: Negotiation has more problems than traditional tendering, because there is not an obvious audit trail of what six people have bid; the cheapest. You have got to find a more innovative way of actually proving that you are going to get value; a whole new mechanism, and it has to be done after the project is finished. Selection has got to be based on trust, in the first place. And at the stage when you are actually tendering, you cannot prove that you have chosen the best contractor.
- SC1: Sometimes, designers take a risk by not charging us for some design work. They may come to the odd meeting and not charge for it, but in truth they will probably charge for it against another job and we are not aware of it. Usually, if they actually put pen to paper or do calculations for us, then it will trigger an invoice.
- SC2: We normally compete as 1:6 but try to do 1:4. We always put in a proper price if we bid. If we do not want the job we tell the client or, if he specially wants us to bid, we price it with a good margin but one which we would work for if successful. We never take cover prices and they are rare. Even if you take a cover price, bidding still has a cost because you must fill in all the paperwork. The new competition legislation may render cover prices illegal with high penalties.

3.5.11 Problems - Monitoring

For the monitoring stage in the commercial process which procurement method and/or selection method seems to cause the most problems?

- CL2: One problem is the inefficiencies of some consultants. We are dependent upon their integrity and capability. They may not be quite as focused or commercial as you would expect them to be.
- MC4: The source of monitoring problems is the documentation you get with a traditionally procured job. It is really geared to create the monitoring. And that is the source of most of the problems.

3.5.12 Problems - Dispute resolution

For the enforcing stage in the commercial process which procurement method and/or selection method seems to cause the most problems?

- CL3: The lowest bidding contractors cause the biggest problems because they always need to recover from a low figure, which is the common problem every client faces I think.
- MC3: We have very few disputes. Currently involved in two; one a bad debt, the other a very late supply of information. Both of these are general contracting. However, most of our work is design and build, where we do have

more control over the design team. Disagreements are resolved very early in the process, so disputes are rare. If the architects are not producing, we can get more resource. There may be an argument with the architect about who is going to pay for it, but at least you get some detailing resource. It is far cheaper than overrunning, which can incur £10-15k a week liquidated damages; you can buy a lot of detailing for that, if you get on with it.

- MC4: In strategic partnering, payment presents fewer problems.
- CO4: Dispute resolution involves a huge amount of management time and a great deal of personal stress.

3.5.13 Other points

- CL1: When the sub-contract chain is long, it is almost impossible to get a main contractor to fix things in snagging.
- CL3: Partnering, and two-stage tendering, are good when people are used to it, but it can take a long time to get people to commit to the new thinking processes.
- CO1: In both contracting and in consulting businesses, the pressure to reduce prices makes it difficult to maintain and train a workforce. Continuity of employment is vital for training consultants, as well. Without continuity, high salaries are needed to recruit people in the short-term, but when these people are laid off, their expectations disrupt the local labour market.

3.5.14 Anecdotes

In your answers to our questions, if you think of any horror stories or anecdotes, please interrupt and contribute them!

- CL1: I have had architects refusing to sign up to even quite simple things like the co-ordination of design. In fact, on one project, a major firm of architects came to see me about being appointed for one of our projects. As I was going through the schedule of duties with him, he made it clear that his firm "did not do that" again and again as went through different items. In the end, I closed my papers and told him that there had been a mistake and that we had chosen the wrong architects. He came round then. But he had been claiming that his PI policy would not cover him for doing even the most usual things for architects.
- CL1: I feel strongly that centralized buying leads to the buyer having a power that is easily abused in some circumstances.
- CO1: Certain clients want to generate information for themselves for free and to do this they invite expressions of interest, and they might issue quite a basic design guide as to what they are after, and they invite quite skilled people who

respond with suggestions. These ideas are then used by the client without the people who suggested them. For example, one local authority had shortlisted 37 people for a £150,000 fee job and invited them all to the same meeting. Clients should be going to the consultants to develop their ideas before they go out and tender it and actually be quite certain that if they are going to tender it, then somebody is going to win. There should be some commitment right from the beginning that they are going to let this as a private service.

- MC3: We often get the foolish situation where the client's design team give us a list of four to six named M&E specialist contractors. When we contact these people, not only have they not had any information sent to them, but they have not been told that they are on a named list, so they are too busy, or not interested, or even don't like working for those people. We then have to go back to the client's team and tell them that out of these six people, two of them can't do it or aren't interested so they then have to give us some other names. It is a complete waste of time, effort and money because people have not bothered with checking that these firms are interested. Another example is when we receive a letter saying, "we have pleasure in confirming that you are on the tender list. Please find enclosed the following documents to enable you to prepare your tender" and a list of, say, 10 items with a note that items 2, 7, 9 and 10 will follow in the next few days. But they have contacted their client to confirm that the letters have been sent out and you then find that within the four week tender period, you might not get all the information until week two or three, so it is not a four week tender period at all, and that just puts more pressure on everyone. These problems seem to arise from the clients' agents agreeing a target date, with the client, on which they are going to tender. If they foul up along the way, they end up rushing things. They want to meet that tender date, so they issue the documentation as it is and that is why we get all these problems because they have not had time to go through it properly, and contact all these people and actually produce documents that are worth having.
- MC3: Where partnering can be useful is in sharing savings between the contractor and the client. So it is in the contractor's interest to keep looking for savings. The trouble is, a lot of so-called partnering agreements tend to be rather one-sided, where the client assumes that, because they have negotiated a job or called it a partnering exercise, then the vast majority of savings goes to them. And if it is a 90% to 10% split, there is no encouragement for the contractor to put a lot of effort into trying to save £10,000 when he is only going to get a £1,000 out of it. Not only does he merely get a thousand, but also it may have actually cost him most of that thousand in general administration and management costs to actually get to that end. If he gets all of that cost back, plus his thousand, then perhaps it is worth doing it. But I think for a partnering situation to work, it has got to be pretty even shares.
- MC3: There is often too much emphasis on how a building looks, sometimes to the detriment of its function. What people must remember is that all buildings are put there for a functional reason. For example, a hospital can be designed

for anywhere in the country, and all you have got to do is to sit it into its background aesthetically. But what I am really saying is the ability to provide true continuity and partnering already exists within the NHS if they really wanted to embrace it. But that is not how it is all written. You need continuity because that is how economies of scale work. That is how the car industry has moved. They used to have 5,000 suppliers, but now they have 500 or even perhaps 50 suppliers.

- MC4: We were involved in a school project. The client budget was not enough for building what was on the drawings. They appointed us as a one-off partner early in the design process. We had workshops with the client to prioritize his needs and we managed to meet them all; we changed the design, did some value engineering, got involved with the supply chain and ended up with a job that went ahead, came in on budget and met or beat all of the targets. However, half way through that process we had a meeting where we presented the re-engineered the job to the whole team; purse holders from the local authority, the schools themselves, the people who did the original design to secure the money, and they were all happy with the proposal except for the architect who said, "hang on a minute not on this company's paper you don't. This practice is not interested in the design you have just come up with. I will not put that on my paper. If that is the sort of scheme you want to draw, get some other architect to draw it for you because it ain't going in my brochure." He was not interested in anyone else's agenda and priorities, he had his own and they were more important than any others. The client found some more money to add back into the scheme, to preserve some elements of the design that this architect wanted, just to keep him on board.
- MC4: We spent four or five years partnering with one client, trying to penetrate the supply chain. For about three of those years we tried to get supply chain deals going with the 45 trades involved and most of them failed: people weren't interested. They wouldn't do it. Either didn't believe in it or couldn't see the benefit for them. They can pretend they are interested but in reality they are sub-letting to someone else who sub-lets it to someone else and by the time you get down to the people who are physically on site doing the work they don't even understand what the question was. So the intermediaries lose interest in it.

3.6 SUMMARY OF STRUCTURED INTERVIEWS

The structured interviews revealed the following general points:

- Generally, people in construction are no different from those in any other sector; construction has its fair share of rogues. You have to be careful about who you deal with.
- Generally, everyone in this business finds that the most effective form of marketing is word of mouth and repeat business.

- While many people prefer negotiation with firms they can trust, when tendering they get between three and six firms to bid, although it can be difficult to get serious bids from this many bidders.
- Everyone prefers to be the only player in their market. Nobody likes to be one of a large number of tenderers, but most acknowledge that some competition is inevitable and acceptable.
- There are many ways of finding a price without competition. Some involve knowledge of the current market rates, others involve cost-plus contracting. In all cases, contractors need intimate knowledge of their own costs.
- Technically, there are many different ways of controlling work on site. The responses to these questions seem to indicate some difficulty in understanding why the questions were being asked, since much monitoring is simply routine management of the job.
- Disputes seem to be rare these days, especially involving contractors. Partnering has played a significant role in the reduction of disputes. Reforms to the legal processes have resulted in fewer litigious episodes.
- Design competitions should not be based on full designs. Partnering seems to be good for getting contractors in early, but it involves a serious commitment and involved negotiations in order to set it up.
- There are some interesting and diverse problems in commercial processes, which are difficult to generalize about.
- Consultants sometimes may cause problems, as can contract documentation.
- Because disputes are rare, they are relatively unimportant in terms of the day-to-day business of construction. When they crop up they can be very problematic.
- The anecdotes provided by interviewees offer interesting and useful insights.

3.7 STAFFING LEVELS SURVEY

The survey design was focused on the construction firm, rather than the project. By asking firms what proportion of last year's turnover was attributable to collaborative working relationships, and then asking how much of last year's turnover was spent on different aspects of procurement costs, two things were immediately apparent. First, firms know how much resource is devoted to particular areas of their business on an annual basis; because they know how many staff they have working on these things and how much they have paid consultants, even if they cannot disentangle these overheads into specific projects. Second, collecting data about a firm's annual activities may reveal nothing about individual projects, but statistical analysis should show whether those who engage in a lot of collaborative work experience higher or lower costs of procurement.

Table 5: Respondents to the survey

Type of participant	Number
Design consultant	28
Specialist trade contractor	20
Main contractor	16
Trade contractor	5
Supplier of bespoke components	4
Public sector client	3
Private sector client	3
Advisory consultant	3
Developer	2
PFI/PPP SPV	2
Other	2
User	1
Supplier of materials	1
Total	90

Table 6: Annual construction turnover

Turnover	£000
Total	5,190,067
Average	57,667
Minimum	115
Maximum	806,300

3.7.1 Distribution

The survey was distributed to named individuals, who were not randomly selected. While these responses are interesting and informative, the lack of randomness means conclusions cannot be drawn about the statistical significance of the data.

3.7.2 Responses

The survey generated 90 responses from throughout the supply chain. Table 5 shows that the single biggest category was design consultants, but about half of the responses were from contractors and suppliers. The whole supply chain is represented here. Table 5 lists the options for how people described their main area of business. Grouping similar participants together, there are 31 consultants, 16 main contractors, 25 trade contractors, 5 suppliers, 10 clients and 3 others/users.

3.7.3 Annual turnover

Respondents were asked for the annual construction turnover of their particular business unit. Table 6 shows that the survey accounts for about £5bn of

Table 7: Collaborative working practice

Percentage band	No. of respondents
Blank	13
10	19
20	11
30	12
40	6
50	6
60	6
70	3
80	6
90	2
100	6

construction work, in organizations whose annual turnover ranges between £115,000 and £806m. The average for all the respondents is £58m.

3.7.4 Collaborative working practices

Respondents were asked what proportion of their turnover could be categorized as collaborative working practices. These responses have been banded together into 10% intervals in Table 7. This shows a preponderance of companies working collaboratively for small proportions of their turnover, with fewer companies having most of their turnover attributed to collaborative working practices

To test for patterns in collaborative working, the amount of work estimated was plotted against size of turnover, but no obvious relationship was revealed. In other words, the chances of finding collaborative working practices are no greater among small companies than large.

3.7.5 Value of construction work bid

One of the questions on the survey sought to discover how much work was bid for during the period. This proved problematic for three reasons. First, the work bid for during the data collection period effectively deals with a period different from that for turnover, as the work that would be carried out during a particular year would have been bid for at various points in previous years. Second, the work bid for in a twelve-month period would not necessarily be carried out in a twelve-month period. And third, many consultants, while indicating the amount of fee income they earned in a particular year, then provided the total construction budget that they had bid for, rather than the amount of fee income they bid for. These results are not reported, due to these inherent problems in their interpretation. As with much of this data, the attempt to measure some useful indicator of the nature and scale of the work revealed more about the problems of measuring contractors' and consultants' performance, than it did about the nature of their work.

Table 8: Construction output by procurement method

Procurement method	Amount	%
Design-build (pure)	£863,099,810	20%
Design-build (novated)	£1,101,205,275	26%
General contracting	£1,584,080,930	37%
Management contracting	£229,488,535	5%
Construction management	£515,766,540	12%
Non-project supplies	£19,342,930	0%
Other	£962,500	0%
Total construction turnover attributed	£4,313,946,520	

Table 9: Favoured procurement

Type	Clients	Consultants	Contractors	Trade cont	Suppliers	Total
Const man	1	5				6
Des-build	1	1	3			5
Direct		2		3	1	6
Fax					2	2
Negotiation	3	3	5	10		21
Partnering	4	3	3	3	1	14
PFI		2				2
Traditional		12				12
Two-stage	1	1	5	5		12
Other	1	4		4	1	10
Total	11	33	16	25	5	90

3.7.6 Amount of work by procurement method

Table 8 shows that most of the projects represented here are design-build variants, followed by general contracting. However, the majority of the design-build work involves novated design teams, a process that can be viewed as closer to general contracting than to pure design-build. There is only a small amount of work taking place through other procurement methods. The difference between the total in Table 8 and the total return of over £5bn shown in Table 6 is due to the fact that not all the respondents attributed their turnover to types of procurement method.

3.7.7 Most and least favoured methods of procurement

The open questions about the most and least favoured methods of procurement have produced a long and diverse list of responses. It is particularly interesting to see how people from different points in the supply chain deal with these questions. For example, suppliers of components tended to respond in terms of how they

Table 10: Least-favoured procurement

	Clients	Consultants	Contractors	Trade cont	Suppliers	Total
Const man	1	3	1			5
Des-build	3	12				15
Dom sub-cont		1		2		3
Dutch auction				2	1	3
Man contract		2	1			3
Measured term		1				1
Novated DB	1	3	4			8
Open comp	3	4	6	12		25
Partnering		1				1
Selective tend			1	1		2
Traditional	2		1	2		5
Two-stage		2	1			3
Verbal order				1	3	4
Other	1	4	1	5	1	12
Total	11	33	16	25	5	90

procure their supplies, with answer such as "buying from UK suppliers", or "buying from trusted suppliers". But not all of them interpreted it in this way, as some stated that they prefer faxed orders and disliked telephone orders. Effectively, these are answers to different questions, because the perception of the question is not common across or within groups of respondents. Similarly, some respondents preferred simple things such as "negotiation" or "two-stage tendering", whereas others preferred complex combinations such as "traditional, in a partnering environment, without competition". The number of mentions for the more common responses is grouped by the main types of respondent in Table 9.

Table 10 shows the summary of responses to the question about the form of procurement least favoured by the respondents. It is interesting to note that similar numbers were both for and against construction management. Open competition is the least popular form of procurement, closely followed by design and build. But while no one favours open competition, apart from suppliers, design-build is unpopular with consultants.

Figure 2 shows the average commercial costs for each stage of the commercial process for the various types of respondent. The graph shows the percentage of

Table 11: Proportion of annual turnover attributed to commercial processes

Type	Selling (%)	Buying (%)
Developer	0.00	0.43
Public sector client	0.44	1.68
Private sector client	0.00	0.57
PFI/PPP SPV	5.63	0.17
User	2.70	0.28
Main contractor	2.57	1.16
Trade contractor	5.43	1.66
Specialist trade contractor	4.48	1.20
Supplier of bespoke components	8.93	2.11

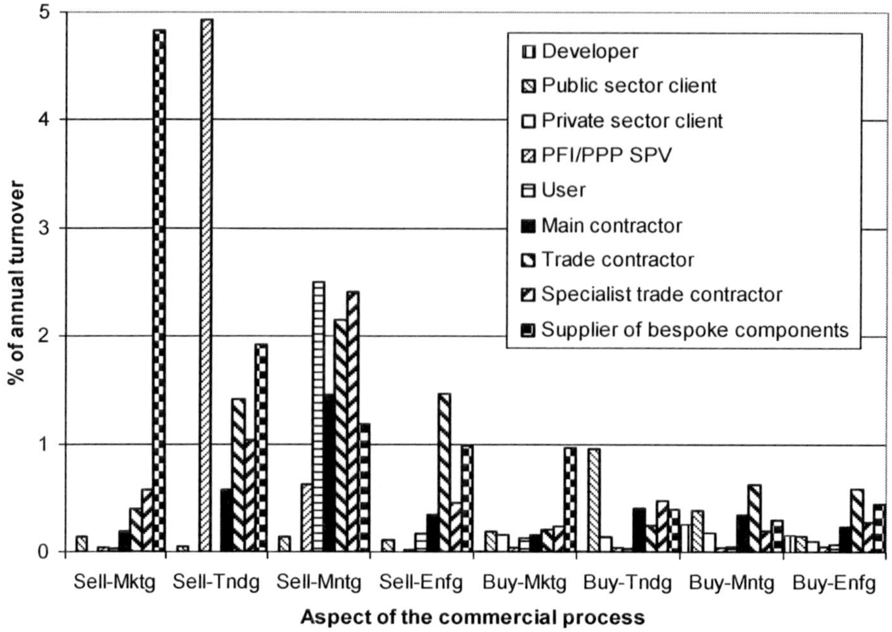

Figure 2: Commercial costs as proportion of annual turnover

annual turnover spent on each stage of the commercial process. The four stages of marketing, tendering, monitoring and enforcing are shown for selling (S) and for buying (B). Suppliers of bespoke components spend about 5% of their annual turnover on marketing, and PFI/PPP SPVs spend a similar proportion on tendering. Much less is spent, in general, on the activities associated with buying.

The primary purpose of this work is to examine whether different ways of working cause any systematic differences in these costs. Thus, the amount spent annually, as a proportion of turnover, has been plotted against the volume of work undertaken using collaborative working approaches. This scatter plot, in Figure 3, shows little discernable pattern. In other words, the expenditure on the commercial processes of tendering, monitoring and enforcing contracts seems unconnected to how much a firm is involved in collaborative working practices.

Table 11 shows the same data as Figure 2, but with types of costs associated with procurement added together. This shows that suppliers seem to spend a much larger proportion of their turnover on the commercial process than anyone else in the supply chain. The price added in a supply chain is interesting. It seems that for each item in the building, the transfer from the supplier to the trade contractor adds 9% to its price, the transfer from the trade contractor to the contractor, 5% and from the contractor to the client, 3%. Even without allowing for any overheads or profit, the simple fact of the existence of the shortest possible supply chain adds around 18% to costs, just to deal with the buying and selling of goods and services.

But if there are several more layers in the supply chain, say, four levels of sub-contracting, this amount could easily be 30%. However, it must be remembered that the alternative (in-house work instead of outsourcing), even if it were possible, would still consume costs: they would just be the costs of recruitment, employment, supervision, monitoring and lack of continuity in work flow for employees.

Figure 3 shows a scatter plot of those responses that included proportions both for collaborative working and for spending on the commercial process. For each point, the eight elements of spending were added together for the Y axis, and this number is plotted against the proportion of work undertaken collaboratively, leaving it to respondents to interpret what was meant by collaborative working. There is no obvious relationship in the graph, so statistical tests were also applied, to seek out relationships. To test for the existence or strength of such a relationship, the correlation coefficient was calculated between the two sets of data. The result of -0.12 indicates that there is a very slight correlation, in that a very few cases of reduced procurement costs may be explained by the adoption of collaborative practices. The lack of a relationship may be because there is none, or it may be due to the wording of the question, which leaves respondents to decide for themselves what is meant by collaborative working.

To investigate further, correlation was calculated for each of the four types of cost. No significant relationship was discovered, values of -0.10, -0.16, -0.09 and

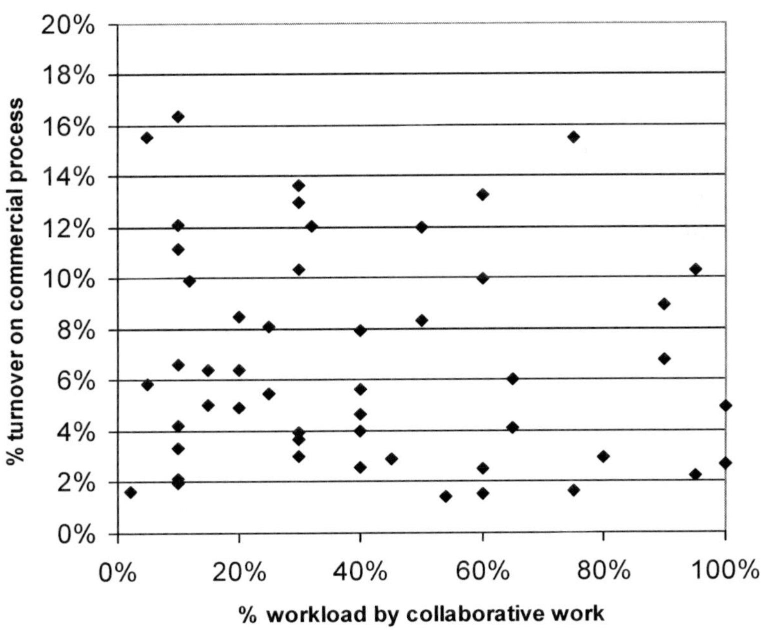

Figure 3: Relationship between costs of procurement and collaborative working

-0.07 arising respectively for collaboration correlated with marketing, tendering, monitoring and enforcing, respectively. In other words, these numbers have no statistical significance, and therefore, the incidence of collaborative working practices is not associated with either higher or lower costs in any aspect of the commercial processes.

Finally, a few questions on the survey form related to statements asking for attitudes to certain issues. Briefly, these revealed that there was marginal agreement with the idea that most people in the industry can be trusted; overwhelming agreement that collaboration is better than competition; definite benefits from adopting new procurement practices; and a feeling that spending on the commercial processes is just about right, with only a small proportion rating it as too much.

3.7.8 Findings from the staffing levels survey

The costs associated with the commercial processes in construction vary between negligible and 9%, depending upon the position in the supply chain and nature of work being carried out. At each step in the supply chain, the costs are cumulatively added, but there is no reason to suppose that procuring things in a different way would eliminate these costs: they may differ, but they would merely be transferred to a different cost heading. Even in a situation of multi-layered sub-contracting, the value-added by each successive party in the supply chain may be only local knowledge and access to various kinds of vendor. But this kind of knowledge is difficult to get any other way, and such qualitative reasoning may be lost in a purely quantitative approach to the question.

From the data presented here, there appears to be no relationship between the level of collaborative working methods and the costs of tendering. While it can be expensive to get into framework deals and partnering arrangements, the expectation of the parties is that this up-front investment results in lower downstream costs. But there is no evidence to support either of these assertions, which is very interesting. This means that there are more influences on these costs than the mere presence or absence of collaborative working methods. However, it must be pointed out that this study is only about costs, not about the benefits of such working practices. It is also important to consider re-phrasing the question about collaborative working practices, as this means different things to different people. It may be better to ask instead about the absence of competition, which is a much easier concept.

The distribution of survey forms needs to be carefully controlled, so that reliable calculations can be made about confidence limits. While the work reported here forms only a pilot study, the final distribution would have to use random sampling. It is interesting how responses to surveys of this nature seem much easier to get from consultants than from any other part of the supply chain. Clients of the industry are the most difficult participants to sample from. This work shows that the methods developed for examining this question are fruitful

Table 12: Average bid cost as a proportion of the value of the work to the bidder

	Full proposal		Pre-qualification	
Procurement route	**Consultants**	**Contractors**	**Consultants**	**Contractors**
General contracting	5.07%	0.81%	3.42%	0.36%
Management	3.06%	0.25%		0.05%
Novated D&B	4.21%	0.21%		
Pure D&B	0.47%	0.80%		1.05%
All procurement routes	4.44%	0.64%	3.42%	0.63%

and worth pursuing. The next stage of the work will be to use random sampling to deal with a much larger number of participants, and to connect the statistical findings with data from interviews.

3.8 BID COST SURVEY

As part of the research, a survey was carried out in collaboration with *Marketing Works* that collected data about 74 bids and pre-qualifications from contractors and consultants. The data reveal several interesting things about tendering practices, and raise some provocative questions.

Table 12 shows the average cost of bids, represented as a percentage of how much work the bidder was seeking to win. It is clear from this that consultants spend a much larger proportion of their turnover in trying to win work than contractors do. Is this because consultants are inefficient? Definitely not. It merely reflects the fact that they are bidding for consultancy fees, not for the whole project. But it is interesting to see that for consultants, general contracting projects are more expensive to win than other forms of procurement. For contractors, the proportion of the contract sum spent on winning work is very much smaller, almost trivial. But there are big variations between procurement methods, in that general contracting, for example, is four times more expensive to bid for than management-based methods or novated design-build.

Table 13 shows how the total bid cost is spread among the various stages in bid management, separating consultants from contractors and lost bids from winning ones. The consultants seem to be more successful when they devote a larger proportion of their energies to marketing and pre-bid discussions, whereas the opposite is true for contractors. Does this mean that contractors are no good at marketing? Perhaps, but another explanation might be that marketing is not something that can successfully be carried out for the purposes of one project, in which case, the successful contractors would not attribute marketing costs to particular bids. In all cases, those who won their bids spent a larger proportion of their bid costs on bid management than those who lost. The consultants who lost bids paid a disproportionate amount of attention to proposal preparation, but there is barely a difference for contractors. Finally, consultants who spend more effort on presentations lose more bids, whereas the opposite is true of contractors. This

Table 13: Proportion of hours spent on each stage of a bid

	Consultants		Contractors	
	Lost	**Won**	**Lost**	**Won**
Marketing and pre-bid	15%	29%	22%	14%
Bid/no-bid	3%	3%	7%	3%
Bid mgmt	14%	21%	18%	31%
Proposal prep	48%	38%	27%	26%
Presentation	19%	9%	11%	15%
Review	2%	1%	3%	3%

Table 14: Averages of responses

Average	Consultants	Contractors
Win ratio	20%	20%
No. of bids per year	72	164
No. of pre-quals per year	93	59
Project value	£1m	£14m
Hours per bid	134	543
Hours per pre-qual	72	474
Cost per bid	£11k	£25k
Cost per pre-qual	£5k	£23k

may mean that bids are not won in the presentations: by the time you get to the presentations, the bid is already won or lost.

Table 14 shows some interesting statistics profiling the overall bid costs gleaned from these 74 bids, with a view to characterizing a typical bid situation (if there is such a thing). Interestingly, the hit rate for consultants and contractors is the same. This means that, whatever the bid costs for a particular project, with an average win rate of one in five, consultants spend approximately 8% of their fee turnover (£16k per £1m of fees) on winning work and contractors spend approximately 3½% of their turnover (£48k per £14m) on winning work. This money can only be recouped from successful bids and it represents a significant proportion of a construction budget. Also, the average fee value for a consultant in the survey is £1m, whereas the average contract value for a contractor is about £14m. And in order to win the average contract, a consultant will spend 134 hours putting a bid together, whereas a contractor will spend only three or four times that amount of time to win 14 times the value of work.

Overall, it can be concluded that marketing should not be left to a project-by-project basis. Consultants are spending a large proportion of their turnover on winning work. The cost of competitive tendering accounts for around 3% of the total construction budget. However, these figures are from a small survey, and cannot necessarily be taken as representative of the whole construction sector.

Because there is so much variability in the responses to the survey, an attempt was made to isolate some of the variables, selecting only main contractors, non-PFI jobs, full proposals only, and these are all shown in Table 15. These bid costs still vary by a factor of more than ten, from 0.07% to 0.80%. However, the biggest costs are associated with a pure D&B job, so in terms of proposals without a full

Table 15: Average bid costs for main contractors in non-PFI full proposals

Sector	Procurement route	Type of bid	Project value	Bid cost
Private	Novated D&B	Partnering	£15,000,000	0.12%
Public	Traditional	Preferred / Framework	£15,000,000	0.14%
Private	Novated D&B	Single stage tendering	£1,600,000	0.33%
Private	Novated D&B	Single stage tendering	£1,600,000	0.33%
Private	Novated D&B	Single stage tendering	£220,000,000	0.11%
Private	Traditional	Single stage tendering	£9,000,000	0.07%
Private	Traditional	Single stage tendering	£3,300,000	0.15%
Private	Traditional	Single stage tendering	£7,000,000	0.11%
Public	Traditional	Single stage tendering	£200,000	0.39%
Private	Pure D&B	Two stage tendering	£250,000	0.80%
Private	Traditional	Two stage tendering	£4,000,000	0.20%
Private	Traditional	Two stage tendering	£600,000	0.57%
Average (of summed values)				**0.12%**

design, the range is still 0.07 to 0.57%. Another thing that is clear from Table 15 is that the partnered job and the framework job are near the lower end of the range, but not remarkably different from the general spread. It is also clear from this Table that larger projects are not necessarily more expensive to bid than smaller projects. Indeed, there is almost no discernable pattern in these numbers.

3.9 SUPPLY CHAIN MAPS

Several case studies have been carried out, and supply chain maps produced. The aim of the case studies is to map a simple construction procurement supply chain. The intention is to identify key supply chain participants and map the chain of their procurement relationships in a typical construction project. The objective is to identify the structure of the chain in terms of the number of tiers of sub-contracting as well as the number of organizations involved in procuring the project.

The case study focused on answering the following five questions:

- How many different work packages occur along the supply chain and how many organizations operate within the work packages? What were the firm's roles, in terms of their function, scope, technology and/or work package?
- What was the number of sub-contractors who quoted for each package?
- What was the basis of appointment in terms of procurement, e.g. design, supply and fix, supply only, labour only?
- What was the basis of selection/competition, e.g. negotiated, tendered, framework, partnering?
- What was the approximate price of the contract/package?

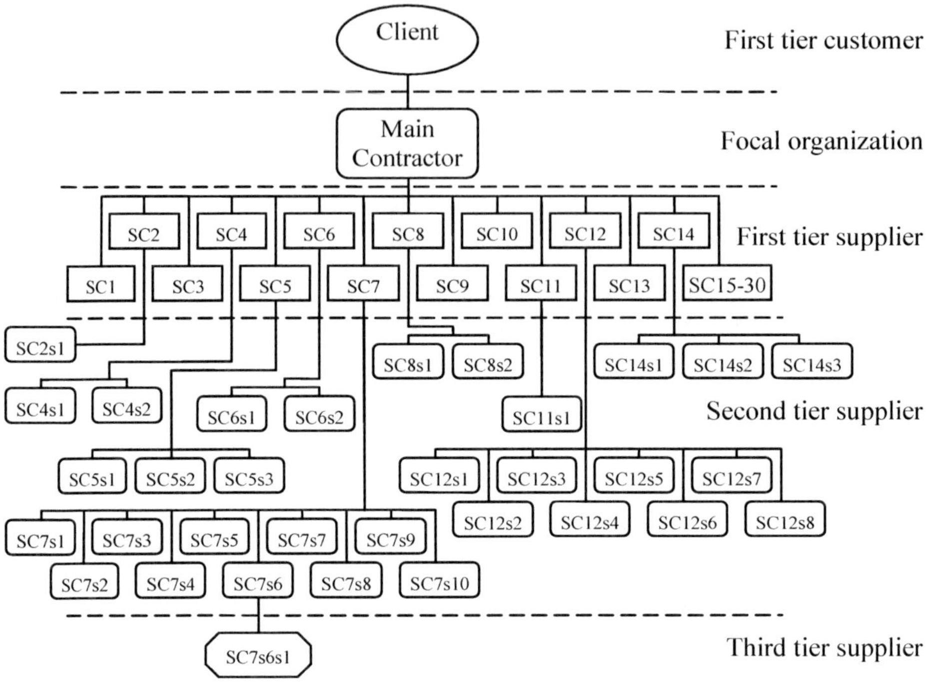

Figure 4: Map of supply chain for case study S1

3.9.1 Case study method

The case study approach involved surveys and interviews in order to map the vertical and horizontal structure of the procurement relationships within specific supply chains. The process for identifying the members of the selected supply chain began with the main contractor (someone with the widest possible knowledge of the project) providing a list of key organizations from whom goods and services were procured in the form of work packages. This included the contact address and telephone numbers of "key contacts". The main contractor then wrote to all contacts named seeking their assistance with the research.

Each "key contact" or organization was telephoned to seek its participation in the study. A mail survey was then sent to each "key contact" mentioned. Each mailing included a covering letter, a summary of the project and the survey instrument, following Dillman's Total Design Method (Dillman 2000). The survey instrument consisted of a two-page questionnaire consisting of six questions. The survey questions were designed to reflect the case study research questions. New

Table 16: Basis of selection of suppliers by immediate customer for S1 project

	First tier customer	**First tier supplier**	**Second tier supplier**	**Third tier supplier**
Negotiated	1	1	14	1
Tendered	-	13	8	-
Framework	-	-	1	-
Partnering	-	-	2	-
Sole supplier	-	-	6	-
Nominated	-	-	1	-
Total	1	14	32	1

mailing lists were compiled from completed questionnaires from active "key contacts". Reminders were sent fortnightly through emails requesting completed forms. In some cases the contacts no longer worked at the company or the company agreed to participate over the telephone but never returned the completed questionnaire. The chain was limited to where there were no sub-contracted work packages or the work was done in-house by the organization or it was a sole supplier or distributor. In effect, information about the supply chain is gathered by each new contact giving information about the next contact. The adjusted sample size, number of respondents and response rates are tabulated.

A structural sociogram was then developed to map the organizations from which services and goods were procured either directly or through a third party. Sociograms can be complex; only key members of each sub-supply chain grouping were used, in order to simplify the mapping.

3.9.2 Case study S1

This project was a £2.6m, two-storey office block and associated car parking and access for a property developer. The project duration was 40 weeks and the procurement route was design and build.

The questionnaire was sent to all 30 contacts/organizations provided by the main contractor as representation of the works packages, functions or specialization for its immediate key suppliers. In this case study the main contractor is considered as the focal organization (Lambert *et al.* 1998). The number of completed questionnaires for analysis and mapping amounted to 14 from these first tier suppliers (see Table 16). This generated 32 contacts/organizations for the second tier suppliers and one from a third tier supplier. The project had one client organization. Overall, 65 organizations were identified as members of the supply chain as shown in Figure 4. However, it must be noted that this number could increase if all 30 organizations and all their key suppliers were to respond. Fourteen sub-contractors responded from a total of 30. It must be emphasized that the end of the supply chain was deemed to occur where a correspondent indicated that all work was done in-house or it was the sole distributor or supplier, such that none of the package was sub-let.

Table 17: Number of firms sampled from each tier for case study S1

	First tier customer	**Focal organization**	**First tier supplier**	**Second tier supplier***	**Third tier supplier**
Number of firms	1	1	30	32	1
Number of firms that tendered or quoted for packages	-	4	103	53	-

**based on 14 first tier suppliers*

Table 18: Degree of sub-contracting by contract value in S1 project

Sub-contractor	**Percentage of MC Package bought from SC (Total MC value £2.6m)**	**Percentage of SC Package sub-let (by 9 sub-contractors)**
SC 1	1.0	-
SC 2	0.9	67
SC 3	1.0	-
SC 4	0.4	50
SC 5	5.4	58
SC 6	0.7	65
SC 7	0.2	-
SC 8	2.2	-
SC 9	0.1	37
SC 10	17.7	82
SC 11	7.3	38
SC 12	0.2	-
SC 13	1.5	88
SC 14	0.4	73
Total	39.0	67

3.9.3 Results and analysis for case study S1

Table 17 indicates the number of identified firms involved in each tier of the chain, compared to the number of suppliers who tendered or presented a quotation for the packages in the various tiers. It indicates that a large number of sub-contractors and sub-sub-contractors exist within the first and second supplier tiers as a result of the many different works packages procured by the main contractor. Table 17 also shows that for the 30 works packages procured by the main contractor, a total of 103 firms tendered in one form or another for the various packages. Each works package attracting between one and seven tenders from sub-contractors. Table 17 also indicates that 32 works packages were procured by the 14 first tier suppliers who responded to the survey and a total of 53 firms tendered for these 32 works packages. Each works package attracting between one and three tenders. Interestingly two of the first tier suppliers procured ten and eight different works packages each from the second tier suppliers.

Table 19: Basis of appointment in S1 project

	First tier customer	First tier supplier	Second tier supplier	Third tier supplier
Design supply and Fix	1	6	3	-
Supply and fix only	-	8	2	-
Supply only	-	-	24	1
Labour only	-	-	2	-
Design only	-		1	-
Total	1	14	32	1

Table 18 summarizes the basis of selection used in procuring the several works packages involved in the construction project by the 14 sub-contractors who responded. The table indicates that 13 out of the 14 sub-contractors were selected based on tendering procedures. However, when the 14 sub-contractors were procuring for their packages, they used a variety of selection methods. The most popular being negotiations followed by tendering and procuring from sole suppliers or sole distributors.

Table 18 shows the 14 sub-contractors account for 39% of the total value of the project to the main contractor. Nine sub-contractors did sub-contract portions of their works package, accounting for 35% of the total value of the project to the main contractor. The 32 second tier suppliers account for 67% of the nine sub-contractors' total cost to the main contractor. This represents 24% of the total cost incurred by the focal organization or main contractor. So approximately 11% of the total cost is equivalent to the value added by the nine first tier suppliers to the packages they procured from their suppliers to supply the focal organization.

The sociogram shown in Figure 4 is based on the framework developed for mapping the supply chain and the model in Figure 1. For this study, the main contractor is the focal organization, equivalent to the assembler (London 2000). The vertical structure refers to the number of tiers in the chain and the horizontal structure refers to the number of firms/trade packages involved in a tier. In this model, the vertical structure has one customer's tier and three suppliers' tiers in relation to the focal organization. Figure 4 indicates that the focal organization sub-contracted 30 works packages to its first tier suppliers (sub-contractors). Of the 14 sub-contractors who responded, nine sub-contracted portions of their works package to the second tier suppliers. In total, the nine sub-contractors procured 32 works packages from the second tier suppliers. The first and second tiers of suppliers in this model indicate that there are potentially a large number of different trade packages resulting in a correspondingly large number of sub-contractors and sub-sub-contractors.

Table 19 indicates that the main contractor appointed the sub-contractors on the basis of either design, supply and fix, or supply and fix only. However, most of the sub-contractors were appointed on the basis of supply only.

Third tier suppliers are mainly materials and product suppliers/manufacturers who supply the sub-contractors and sub-sub-contractors as and when necessary. In other words, these sole distributors/manufacturers keep stocks of materials and products irrespective of ongoing projects and supply them when required by the sub-contractors. The number of third tier suppliers/manufacturers would be at least the same as the number of second tier suppliers who provide materials along with their labour. One of the interviewees for the supply chain case study sourced materials from 34 different manufacturers. The third tier of suppliers can also become complicated if there are middlemen between the manufacturer and the sub-contractors. This can elongate the chain to several more tiers making it complex and difficult to manage. As the third tier suppliers are on the periphery of the construction sector, it would be appropriate not to map the chain beyond them.

In the supply chain, many different selection methods were evident, such as competitive tendering, negotiations, framework agreements and partnering. In this case study, the main contractor prefers competitive tendering for its works packages because the products or services offered are specialized or customized. However, the first tier suppliers or sub-contractors prefer to negotiate their packages with their suppliers in the second or third tier supplier band.

In terms of the basis of appointment, the structure appears to follow a hierarchy of the first tier suppliers being appointed to design-supply-and-fix, while most of the second tier would be appointed to supply-and-fix or supply-only, depending on the goods being procured. Most relationships beyond the third tier are supply-only. In such cases, a form of framework agreement is developed where the products or goods are supplied as and when required by the sub-contractor.

3.9.4 Case study S2

The second case study was a £3.2m refurbishment and conversion of existing hospital wards into research facilities for a medical school. The project duration was 48 weeks. The project was procured using a two-stage tendering process with PC/Works/1 as the main contract.

As with the previous case study, the main contractor was considered as the focal organization. There were 12 contract works packages in the first tier. In addition to these 12 works packages were 18 suppliers from whom the focal organization procured goods directly. The 12 first tier suppliers generated 56 contacts/organizations for the second tier suppliers and 8 from the third tier suppliers. The project had one client organization and five consultant organizations in addition to the supply chain. Overall, 101 organizations were identified as being members of the supply chain as shown in Figure 5. It must also be emphasized that the chain did not go beyond those organizations where all work was done in-house or where it was the sole distributor or supplier such that none of its work was sub-let.

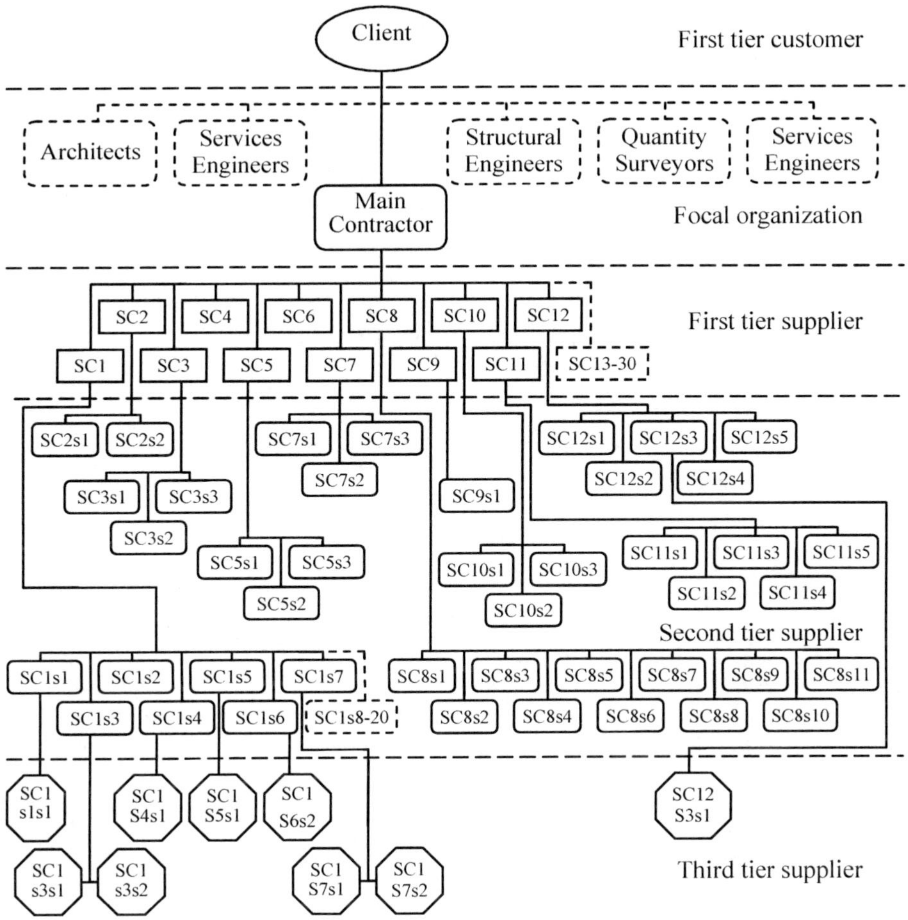

Figure 5: Map of supply chain for case study S2

3.9.5 Results and analysis for case study S2

The totals in Table 20 indicate the number of firms involved in each tier of the chain. It shows the large number of sub-contractors and sub-sub-contractors within the first and second tiers. The supply-only appointments were generally from builders' merchants and not sub-contractors (18 merchant suppliers). Therefore, the main contractor appointed the sub-contractors on the basis of either supply-

Table 20: Basis of appointment within supply chain in S2 project

	First tier customer	First tier supplier	Second tier supplier	Third tier supplier
Design supply and fix	-	-	-	-
Supply and fix only	-	10	10	-
Supply only	-	18	39	9
Labour only	-	2	7	-
Design only	-	-	-	-
Total	-	30	56	9

and-fix, or supply-only. It is also clear that 39 out of 56 second tier suppliers only supplied materials and product, while just ten provided labour services only.

Table 21 shows that the 12 sub-contractors account for 69% of the total value of the project to the main contractor. Ten sub-contractors sub-contracted portions of their works packages, accounting for 35% of the total value of the project to the main contractor. From Table 21, the sub-let rate of the first tier sub-contractors ranges between 33% and 90% of their works package.

Figure 5 indicates that the focal organization sub-contracted 12 works packages to its first tier suppliers and 18 packages to builders' merchants indicated by the broken lines (no further data was available from these suppliers). Of the 12 sub-contractors, 10 sub-contracted portions of their work to the second tier suppliers. The first and second tiers of suppliers in this model indicate that there are potentially a large number of different trade packages resulting in a correspondingly large number of sub-contractors and sub-sub-contractors. These first tier sub-contractors were mainly appointed on supply-and-fix basis while the majority of the second tier sub-contractors were appointed on a supply-only basis. However, it was noted in a particular case that one first tier sub-contractor (SC1) sub-let most of its work to second tier sub-contractors on a supply-and-fix basis.

Table 21: Degree of sub-contracting by contract value in S2 project

Name of sub-contractor	Percentage of MC package bought from SC (Total MC value £3.2m)	Percentage of SC package sub-let % (by 12 sub-contractors)
SC 1	41	40
SC 2	1	46
SC 3	0	45
SC 4	2	-
SC 5	2	90
SC 6	1	-
SC 7	3	59
SC 8	6	89
SC 9	1	33
SC 10	1	-
SC 11	2	55
SC 12	8	90
SC 13- 20	-	-
Total	69	-

It was also evident that sole distributors/manufacturers supply to all of the three tiers depending on the product. For example, in this case study, a sub-sub-contractor supplied fixings and supports to the main contractor as well as to a sub-contractor. The interviews revealed that the sole distributors sourced their products from several manufacturers.

In this case study, the main contractor preferred competitive tendering for its works packages. However the first tier suppliers preferred to negotiate their packages with their suppliers in the second or third tier as a considerable number of these firms supplied goods or services only. Trading partners tended to be distinguished on the price and the delivery dates as well as the charge or pay and how easy they are to work with.

In terms of the basis of appointment, the structure appears to follow a hierarchy of the first tier suppliers being appointed to supply-and-fix, while most of the second tier suppliers were appointed to supply-and-fix or supply-only depending on the goods being procured. Most of the relationships beyond the third tier are supply-only.

The sub-contractor selection process used by the main contractor was in the form of a formal interview and questionnaire by a team of consultants. Each team interviewed an average of six-eight sub-contractors for each works package. This implies the presence of a large number of abortive bidders in the first tier. Issues covered during these interviews included the sub-contractor's supply chain, health and safety record, financial performance and procedures, and human resource management issues. The sub-contractors were scored on a matrix based on the interviews.

The case study interviews revealed that 90% of the sub-contractors engaged were part of the main contractor's supply chain and the remaining 10%, who were specialist sub-contractors, were specified by the designers, who had worked with these specialist sub-contractors previously.

This case study has identified a large number of participants particularly at the second tier. The mapped supply chain can be described as a three-tier supply chain from the sole distributor to the focal organization with a horizontal structure of 12 works packages in its first tier of suppliers and more than 56 trade packages in its second tier. This case study has revealed a large number of abortive bidders.

3.9.6 Findings from supply chain maps

The case studies describe three-tier supply chains from the sole distributor to the focal organization with a horizontal structure of 30 works packages in the first tier of suppliers and more than 32 trade packages in the second tier. This is in contrast to the widespread perception that construction supply chains have long vertical structures or a series of tiers from the contractor to the sole distributor or manufacturer.

The case studies have revealed a large number of participants, particularly at the second-tier. This contrasts with common characterizations of the construction process which identify only a dozen or so organizations. These studies bring to light the reliance, on the part of the main contractor, on a large pool of sub-contractors and sub-sub-contractors to procure the construction project. They also highlight the importance of being able to picture the supply chain structure as well as the large number of firms involved in procuring work for a construction project. This may necessitate a dynamic (re)configuration of supply chains over time to take advantage of better configurations to reduce costs and time, and increase efficiency.

The supply chain maps demonstrate the large number of transactions that take place on relatively straightforward projects. Every one of these transactions involves a tendering process, and this is particularly wasteful if every sub-contractor is asking six-eight third tier suppliers for bids. On case study S2, for example, there must have been approximately 72 tenders prepared by third tier suppliers. This raises a serious question about how much these structures cost to set up. On the other hand, if work was not sub-contracted, but carried out in-house, the internal costs of management and supervision would have to be accounted for, instead of tendering costs. Furthermore, it must also be pointed out that these two examples are relatively small and straightforward projects. More plots of this nature would help in understanding what is typical.

3.10 CONSTRUCTION MARKET DATA

One of the costs of tendering must be the amounts payable to contract-lead companies like Glenigan, and other market intelligence. Not much mention has been made of this as a cost, although Glenigan was mentioned in passing a couple of times by people who were sceptical about the value of their data. However, as part of this research, it was thought at one point that the total costs of tendering to society as a whole might be calculated, and thus Glenigan were approached, who were very willing to collaborate on this research project. Unprecedented access was provided to their database, which covers almost every planning application over the last ten years. While it was not possible to collect sufficient data from the industry to multiply up to an industry-wide account of tendering costs, the data was so useful and powerful that a separate research project about the UK construction market was carried out (Gruneberg and Hughes 2004, 2005).

4 Fresh perspectives on construction procurement

This research has revealed a wide range of insights into the tendering processes in construction. There are costs to all of the participants in the process, and superficial efforts to economize on these costs often seem to add to the costs when the whole picture is viewed.

4.1 COSTS TO PARTICIPANTS

4.1.1 Costs to the client

The decision on how to procure a project is taken by the client. This choice will affect not only the client's costs, but also those of all the parties involved in bidding and may, in the long run, have consequences for the industry and society as a whole.

Client costs are measured mainly in management time but their decisions will put more or less work on their external advisers. In both cases the real cost is the opportunity cost, that is, the value of the output they would be producing if they were not working on that particular project. Opportunity cost can be more or less than money cost. *Prima facie* it would seem that the client's costs would rise with the complexity of the procurement process, though we are not aware of any hard data on this matter. It would seem that extra functions required by the contractor or manager would require more expensive management of the selection, especially if there is competition for each function. Such extra costs would be offset to some extent, or even perhaps more than offset, by the saving on the costs of selecting those performing these functions separately. Many large continuing clients have moved to new methods of selection of which partnering is one. In this case selection of all the parties to the process is undertaken in advance of requirements so that when a particular project is let no further selection should be necessary. The reasons for adopting these diverse methods of procurement do not rest on the costs of procurement but on a belief that efficiency in the total provision of the required project is in the long run enhanced, for example by lower costs or less risk of failure (Hughes *et al.* 1998).

Clients do not take into account, and probably do not realize, that the greater the costs imposed on contractors by their procurement methods, the greater in the long run will be the costs of construction by the process described below.

The client's costs of procurement are private costs; they are under the client's control and it may reasonably be assumed that the advantages bought by the client in incurring these costs yield a financial benefit. Moreover, they are directly related to the individual project and so are variable. An element of fixity is introduced with partnering where costs of getting to the stage of a partnering agreement are related to an unknown number of projects and cannot afterwards be altered.

4.1.2 Costs to the contractor

The building or civil engineering contractor may become involved with provision of services beyond those of construction, for example, of design, of finance for the construction of a project, of management of the completed facility and the services it provides. This involves team working and the constructor may not be the leader of the team. For present purposes, "contractor" is understood to mean the party to the process who is contractually committed to the project of which construction is an important part.

The contractor is not in control of the costs of obtaining work. A contractor who wishes to be in a particular business has to accept the procurement method ordained by the client. There are two aspects of costs of obtaining work to be considered in this section: the cost of winning a contract for one project with various types of procurement methods and the cost to contractors of tendering for work where they failed to obtain the jobs.

On the face of it, the costs of being selected and of estimating will be lower for simple projects, especially when the contractor has knowledge and experience of that type of work. If functions other than construction are added to the bid, notably design, provision of finance or management of the services provided by the project, the costs of bidding escalate. For partnering work the initial costs to the contractor of passing the selection process may be very high. This is especially so since the process will probably take the time and effort of a large number of senior staff for whom the opportunity cost may well be higher than their money cost. Once the contractor has passed the tests, the cost of estimating the price, which for multidisciplinary projects may itself be high, is the only cost. Costs of being vetted for partnering agreements cannot be altered. They are the same whether the firm gets no work as a result or a lot of work. The manpower used up in the process of selection *ex-post* has no alternative use so it has no opportunity cost and ought not to be considered in the decision-making process. This means that the firm should not tender for a project for which it has gained a position on the select list just because it has spent a lot of money and effort in getting to that position. The decision should be based on the current and future costs and benefits only. Costs which have been incurred in the past and have no current opportunity cost are known as sunk costs.

There will also be the problem that, if the contractor is inexperienced, the estimate may be too low so that the actual project makes a loss. Contractors are particularly vulnerable if they rely on outside specialists with attendant risks of delays, escalating costs and non-fulfilment of contracts. These risks are not part of the cost of obtaining work but should rather be included in the contract price as part of the mark-up.

The cost of estimating and the costs of being asked to bid for an individual project are variable costs, that is, they are directly related to turnover, though the success rate determines the exact relationship. The cost of estimating for a particular project could be included in the cost estimate for that project. It is understood that that is not normal practice. All costs of obtaining work are regarded as overheads and included in the mark-up for contracts as appropriate. The total cost of obtaining work is high. Taking the estimates of costs of tendering previously mentioned of ½-1% for traditional contracts and 2-3% for those involving finance, and assuming that for traditional contracts contractors obtain one in six of contracts bid for and one in four for complex projects, then the total costs of obtaining work become 3-6% for traditional work and 8-12% for complex work. The National Joint Consultative Council for the UK recommends five to eight contractors for selective tender (National Joint Consultative Council 1994). It is not clear whether these figures are correct but they indicate that costs are high and that costs for complex work are probably a higher proportion of turnover than for simpler work. Thus the client, in determining what procurement method to use, is affecting contractors' overheads across the board and not simply for the one project being procured.

If the change to more complicated procurement methods increases the cost of obtaining work and hence overheads, can contractors simply add this extra amount to their overheads and hence charge more for contracts generally or will they, in so doing, become uncompetitive and lose their place in the market? There is no evidence of what happens in practice but theory goes some way to elucidating the position. Answers depend, amongst other things, on the level of competition and the relative power of the players in the construction process.

If there is effective competition in the industry, then contractors in general will be making just enough profit to keep them in the industry. This level of profit is known as "normal profit". If they make more than that, others will enter the industry and their extra capacity will drive down prices to the level at which normal profit is made. If firms are making less than normal profit some firms will leave the industry, thus reducing supply so that prices rise to allow normal profit. Individual firms may in the short-run make more or less than normal profit but in a truly competitive situation it will pay them to expand or to shrink so that they too are making normal profit. The economic system is in a constant state of adjustment towards an equilibrium situation (Hillebrandt 2000).

With effective competition, if all contractors were in the same markets in the same proportions, they would all have the same increases in overheads and their costs would rise to the same extent. They would be able to increase their prices but as they did so the demand for construction would fall and some firms would shrink or go out of business. However, all contractors are not in the same market. If those relatively few contractors who are in the complex procurement project market as well as in other markets try to add the extra overheads for the former onto all their projects, they will become uncompetitive in the traditional markets where they are competing against contractors who are operating only in these markets. Their tender success rate will fall, their costs will rise and their profits will fall. However, if they add the higher mark-up on to the complex projects they will be in the same position as other contractors, provided their competitors are doing the same thing. Logically they all have no choice but to separate the markets in applying overheads though it is probable that, in determining mark-ups, this argument is swamped by the other factors determining mark-ups, such as the overall state of the market, and the expected competitors. Nevertheless a rise in price may choke off demand as in the case where all contractors are in the same market.

If competition in the industry is not strong, for example, if there are few contractors in the market, if there is overt or tacit collusion or if one contractor has a monopoly, the profits being made will be higher than normal and the contractors will probably absorb part of the higher overheads so that the price rise is less than in a competitive situation, or even nil.

The construction industry is notoriously subject to fluctuations in demand so that the overall market demand and supply situation is constantly changing. This means that equilibrium is never reached and often, for long periods, not even approached. Thus in the short run the effect of higher costs will depend on the relative power of clients and contractors. If work is short, the contractors will have to bear extra costs and profits will fall below normal levels. In a period of boom, when clients are having difficulty in finding contractors to undertake work, the contractors can recoup all overheads and much more besides. Profits will be super-normal.

It is important to realize that high costs of obtaining work and of pricing projects, through their effect on contractors, affect other clients and the operation of the industry as a whole. Because a healthy industry is important to the economy, that is a matter of concern to society as a whole. Society is also concerned with other broader issues which are discussed below.

4.1.3 Costs to society

There are two main ways in which the method of procurement adopted by clients impacts on society. The first is that the construction industry costs can rise, as outlined above, so that the price of construction increases to the detriment of other clients and the economy. However, there may be advantages of a more elaborate procurement process if this leads to better buildings in terms of, for example, design, quality of construction or life cycle costs. Clients take account of the benefits and their own private costs but not of the increase of costs on the industry. Thus the private costs and the social costs are not the same.

The other, probably more important, consequence of elaborate procurement processes is the use of scarce resources whose cost is not necessarily reflected in their price. Consider, as an example, the resources used in design. In the traditional process for building the client chooses and commissions an architect to prepare the designs for a project. Only one architect is involved. In a design and build competition, or simply in an architectural competition, several architects are producing designs for the same building of which only one will be used. In other types of procurement, such as PFI or prime contracting, many designers are employed on different solutions for the same project. Complex sub-contract packages may also involve detailed design inputs. The money costs of this will be reflected in the costs of the individual contractor and the cost of the unsuccessful bids must be recouped as outlined above. However, the real cost may be quite different. If there is a shortage of good architects, while some of them are working on the one project, they are not working on other projects and, as a consequence, some projects may not get built, may be delayed or be designed by less competent persons. The opportunity cost of employing these architects would be high. In the long run economic forces would correct this situation. The price (fees) of architects would rise and more architects would emerge. The problem is that the training period for an architect is seven years and for an experienced architect much longer. The rise in the fees of architects would serve to increase the cost of projects without solving the supply problem. In this case the social costs of elaborate procurement methods may be much higher than the private costs to the client.

It may be that there is a surplus of excellent architects. In that case the opportunity cost of the employment of several on one project is low and may be lower than the money cost. Their use would be beneficial to society in other ways, perhaps reducing unemployment benefits or social security payments, and certainly, by increasing their incomes, stimulating the economy as a whole. Thus, in this situation, the social cost of employing architects on the project with a complex procurement process may be less than the private cost.

Similar arguments may be applied to the cost of arranging financial packages, assessing and minimizing full life costs, assessing the costs of providing a service linked to the building or other works and so on.

4.2 COMMERCIAL PROCESSES IN THE CONTEXT OF PROCUREMENT METHODS

The data collected illustrates the remarkable diversity of procurement practices, and demonstrates why it is that there is so little systematic data available on the commercial costs of procurement. Given the six variables in Table 1 (page 12), the fact that each of the six procurement variables has four or more options means that the range of combinations is huge. Even if there were only five possibilities under each category, there would be 15,625 (i.e. 5^6) different ways of procuring a project. While it is often said that every project is different, this simple calculation demonstrates that there are so many different possible ways of procuring a project, that even if the technical specification of buildings were the same, there would still be enormous variability in the procurement processes.

This variability explains the huge range and diversity in the projects and firms surveyed, indicating that this kind of research needs to focus on specific project and participant types, if any meaningful numbers are to be elicited. The return rates for the surveys are so low as to make generalizability of the results impossible. But the numbers referred to in the previous paragraph show that a generalized account of the practical processes involved in getting work, and monitoring and enforcing contractual performance is not practical.

One important fact in the staffing levels survey is that while there is a slight correlation between reduced tendering costs and increased collaboration, of -0.12, this means that 88% of the reduced costs of procurement are **not** explained by the increase in collaboration. The difficulties of finding data that will clearly demonstrate a connection between costs of the commercial process and the amount of collaboration is probably due to the lack of any such relationship. There are simply too many variables to enable such a simple cause and effect conclusion.

The research also shows that there are different ways of looking at the same thing. While an early anecdote (section 2.3.1) revealed that there could be many, apparently wasteful, tiers in the supply chain, the conclusions from the staffing levels survey (section 3.7.8) showed that multi-layered sub-contracting can add value, even if it is only local knowledge. Without these multiple tiers, large firms would simply not have access to the kind of knowledge they would need to procure locally. Moreover, while work may be passed to smaller companies, risks are never fully laid off along the supply chain, as ultimately, if a small sub-contractor ceases to trade, the next tier up the contractual chain still carries the responsibility for getting the work done. It is a moot point how much this might be worth, in terms of value, but to the extent that the pricing mechanism of an open and competitive market place is operating, the multi-tiered supply chains are not as horrific as some people believe. And on the supply chains that have been mapped, there was no evidence of unnecessary tiers.

Although the PFI bid studied was very expensive compared to traditional bids, the SPV is expected to prepare a complete design and operational strategy for the 30 year life of the facility, and, therefore, these costs have to be compared to the total of the contractor's and trade contractor's bidding costs as well as the design costs in a traditional system. In this case, the PFI scenario may be cheaper than the traditional scenario; it is merely that the costs are borne by different people that they are so apparent. But more data would be needed for a more general statement about PFI bidding costs.

One interesting thing that has emerged is the cost, for certain D&B contractors, of their involvement in a PFI bid. Again, it depends on the specific arrangements, and they are all unique. But in some cases, a D&B contractor may be bidding for the construction work to an SPV, who in turn is bidding for the concession to design, build, finance and operate a facility over its life-span. In some situations, the SPV is bidding on the basis that this particular D&B contractor will get the job, if the SPV wins. And in this case, the D&B contractor has to pay for the entire design work up-front, at its own risk. This can be a significant proportion of the construction costs, and this is the kind of cost that contractors worry about when they complain of the high costs of the PFI tendering process.

4.3 COMPETITION, COLLABORATION AND POWER

There are some interesting relationships between funders, clients, contractors and sub-contractors. Changing the nature of the competition between them is bound to have an impact on the relative economic power in the bargaining process.

4.3.1 The power of the large continuing client

For the projects which the client has decided to construct, there is only buyer; no-one else will build that project on that site. This is true whatever the situation of the client and, provided that there is a multitude of other clients for the industry, there are no great problems in this situation. However, a client that is a large organization with a high value of projects to be constructed, now and in the future, is extremely powerful and is able to exercise some monopsony power. This power is enhanced by a well-qualified team handling the work. It will be greater if the work has special characteristics and if the client dominates a specific market. The client in this situation may be in a position to dictate the terms of contracts because the work is so important to the contractors in the industry that they are willing to accept unfavourable terms and conditions in order to obtain work. They are fully aware that the client can get contractors elsewhere, including abroad. In the traditional method of competitive tendering this has often led to skimped work, as the contractors try to make some profit, and to profits which are inadequate for the long-term health of the industry.

4.3.2 Collaborative working

Solutions to aspects of this unsatisfactory situation have been sought at intervals over nearly 70 years starting with the Simon Committee Report in 1944 (Ministry of Works 1944). The earlier reports concentrated on changing the tendering system, for example, recommending cessation of open tendering and reduction in number of bidders. In the last few years Latham (1994) and Egan (1998), to mention just the most influential reports, have attempted more fundamental solutions. Recent reports have had a large client input which was rare in earlier ones. The emphasis now is on collaborative working to obtain increases in efficiency and productivity, certainty of cost and delivery time. Many public and quasi-public sector organizations, as well as supermarkets and property developers, have set up schemes to arrange this. The advantages of collaborative working are claimed to be substantial.

It is yet early to determine how far gains have been achieved although there is evidence of some significant advantages. Little attention has been devoted to the costs of the collaborative working. One aspect of the costs of collaborative working compared with other methods has been investigated by this project. At the present time the way in which collaboration takes place is still developing. Some of the processes may have very different effects in the longer term than would appear when they are set up. The method of selection of a contractor for collaborative working can be the same as for traditional arms-length contracting, for example tenders or negotiation. This has been used by a number of clients and would be usual for one-off clients. On the question of selecting contractors on a basis other than price, there is useful guidance published by CIRIA (Jackson-Robbins 1999).

One matter for consideration is that for procurement methods like ProCure21, contractors are thrown together and often share information. This should lead to increased expertise in this particular type of building. It is also an ideal opportunity for the selected group to discuss not only technical matters, but also the terms and conditions of their employment. It could develop into some form of combination limiting competition. How powerful such a group could become depends on a number of factors:

- The terms of the contractual agreement with the client and, in particular, whether the client is legally able to go to other contractors within the allotted period.
- The number of equivalent contractors outside the group.
- The extent to which the group of contractors has developed technical know-how which would not be available to those outside the group.

In any case the group could well arrive at a situation where they act as one, that is as a monopolist for the duration of their contractual arrangement.

4.3.3 Bilateral monopolies and market imbalance

The moves towards collaborative working clearly carry great benefits, and these are articulated clearly enough elsewhere. But there are disadvantages. Continuing clients may become very powerful with a combination of expertise and a large workload. There may be parallels with large retailers who have pushed the prices of British food so low that farming is often no longer a viable business in the UK. The danger is in creating a monopsony, making it difficult for contractors to price their work sufficient to make good returns. On the supply side, it would be very difficult to replace a contractor who develops specialized and detailed technical knowledge for the purposes of providing focused construction expertise to a particular client. The opportunities for developing a monopolistic position would be great, and this may result in an imbalance of market power towards these contractors.

Potentially, the powerful continuing client as a monopsonist could be dealing with a cartel which, for a period of time is a monopsonist. This situation is known as bilateral monopoly. The outcome of deals in this situation is indeterminate depending, for example, on the position of both parties in a broader perspective, on bargaining skill and psychology. The client may have to raise the level of profit for the contractors or perhaps withhold work rather than give in to the cartel. In any case this is confrontation not collaboration.

Collusion is another potential problem, although it must be emphasized that there was no evidence of it in the research. Indeed, it may be that too much reliance on price-based competition may be a more likely precursor to collusion (Dorée, Holmen and Caerteling 2003); traditional assumptions about competitive markets seem to miss the complex realities. In setting up long-term relationships with small numbers of suppliers who are known to each other, is there anything other than their honesty and integrity to guarantee that they will not collude?

4.3.4 The effect on the rest of the industry

If the arrangements of pre-selection of contractors continue to go ahead, the work load of continuing clients will be taken out of the common pool of work for considerable periods of time. In the short run this is unlikely to have a great effect on the rest of the industry. The initial contracts might have gone to these particular contractors in any case. As the selected group continues to have more contracts without the competition being available to the other firms in the industry, the latter will see their prospects for development being restricted. There are now in the industry some contractors who are desperately trying to obtain long contract commitments from continuing clients because they think that, once the groups of contractors are selected, their opportunities are severely restricted for some years ahead. The situation is exacerbated by the fact that so many of these arrangements are being organized at one time, so that the next major phase may be, say, five years ahead. Had they been staggered, as they would have been had the arrangements been in place for a long period, the problem would not have been so acute. As it is, the recent changes are reducing flexibility and may well hamper the development of up-and-coming contractors.

4.3.5 Cover bids and collusion

An invitation to submit a bid may involve a contractor who does not want to do the job. It is widely believed that having been invited to submit a bid, failure to submit a bid would result in no further invitations to bid, effectively being struck off tender lists. It is not clear why this should be so, but the typical response of a contractor who does not wish to win a project for which it has been invited to bid is to submit a "cover price".

Zarkada-Fraser (2000) studies the ethics of collusive tendering, highlighting the problems that might compromise competition or defraud clients. Although collusion may influence the submission of cover prices, which in turn reduce the cost of tendering, there is no quantification of the costs involved. However, it may be deduced from this paper that the high costs associated with traditional tendering practices are part of the temptation to collude and if tender costs were not so high, perhaps contractors would be less inclined to submit cover prices.

5 Conclusions and recommendations for further work

In order to deal with the complex and various methods of procuring construction, it was necessary to develop a structured and rigorous definition of the differences between methods for procurement and selection, resulting in six categories of definition (section 2.2.4):

- Ownership, initiation and funding
- Selection method
- Price basis
- Responsibility for design
- Responsibility for management
- Amount of sub-contracting.

The purpose of the research was to deal with both the costs of buying and the costs of selling to different parties in the construction process, to see if they differ between different ways of working. But as shown in section 4.2, there are perhaps tens of thousands of different combinations of these factors.

5.1 QUANTIFYING THE COSTS OF TENDERING

One of the most significant findings from these attempts at quantifying the costs of tendering is that the number of variables is huge, and it is impossible to isolate factors that influence costs directly. In order to draw some conclusions, a useful solution may be to pick projects that typify common approaches to procurement.

By considering the responses from interviewees, coupled with the returns from the bid cost and staffing levels surveys, an assessment has been made of the relative costs to a contractor at each stage of the commercial process, in selected different procurement processes. This is shown in Table 22 for five typical procurement arrangements.

The amount spent by main contractors on the commercial process varies enormously. This is because some work is let purely based on competition on price, whereas other work is let based on longer-term relationships and/or contractor's design. Some projects involve the contractor marshalling a supply chain in order to even prepare a bid. In traditional contracting, where the contractor prepares a tender based on design documentation and bills of quantity that have been prepared by the client's team, a contractor may spend something in the region of £10-50,000 preparing a tender for projects in the range of £1-50m. In other words, the cost of the tender itself is very small as a proportion of the contract value. On the other hand, the task of finding work and bidding for it is carried out by senior and/or highly specialized staff. Such staff may spend all their time on these activities, and their capacity to find opportunities and bid for them can become the limiting factor on how much work a contractor can take on.

Given that most contractors sub-contract most of their work; these costs of tendering may represent a larger proportion of the main contractor's income. Although every member of the supply chain will also need to prepare tenders in order to take part in the work, these costs cannot simply be removed by bringing work in-house, as the costs of employment may be higher than the costs of tendering, especially if demand fluctuates, or contractor's resources have to be moved to widespread work locations to prevent them from lying idle.

A crucial factor in all these figures of costs of tendering and marketing is that they refer to the cost of one attempt to obtain work. The success rate of tenders, according to the bid cost survey, is about 1 in 5. Therefore, although the cost to the contractor for a single successful bid may be, say, £48,000, the cost overall would be £240,000. The contractor has to re-coup the cost of failed bids in one way or another, and the usual procedure is to regard the costs of marketing and bidding as an overhead, especially as no contractor has the cost-accounting capacity to track these costs on a job-by-job basis. This will be included in the pricing of all projects. In this way, the failed bids add to the cost of construction.

Table 22: Commercial costs to contractor of each stage in different procurement processes

	Stage of commercial process			
	Marketing	**Agree**	**Mgmt**	**Dispute**
General contracting, measured bills(1)	V. low	V. low	Med	Low
General contracting, drawings & spec (2)	V. low	Low	Low	Low
Design and build, pure (3)	Medium	Medium	V. low	V. low
Design and build, novated (4)	Low	Medium	Low	Low
Private Finance Initiative (5)	High	V. high	V. low	Low

NOTES:

1. This could be owner-financed, contractor selected by selective tender for a one-off project, price is for the work and materials as defined by a detailed bill of quantities, independent design responsibility, contractor has overall management responsibility for construction process, high level of sub-contracting.
2. This could be owner-financed, contractor selected by selective tender for a one-off project, price is for the work and materials as defined by a drawings and specifications, independent design responsibility, contractor has overall management responsibility for construction process, high level of sub-contracting.
3. Also called "pure" design and build, this could be owner-financed, contractor selected by competitive tender for a one-off project, price is for a whole building, contractor is responsible for design, contractor has overall management responsibility for construction process, high level of sub-contracting.
4. Also called "pure" design and build, this could be owner-financed, contractor selected by competitive tender for a one-off project, price is for a whole building, contractor is responsible for design but being appointed after the architect has no influence over early design decisions, contractor has overall management responsibility for construction process, high level of sub-contracting.
5. This could be PFI-funded, contractor selected by competitive tender for a one-off project, price is for a whole building, contractor is responsible for design, contractor has overall management responsibility for construction process, high level of sub-contracting.

According to the staffing levels survey, a contractor spends around 2½% of turnover on sales activities. By comparison, specialist and trade contractors spend about double on selling their services than the main contractors. They have higher commercial costs for all types of procurement, though the enforcing costs of specialist and trade contractors are nearer to those of the main contractor. The amount they spend is affected by the type of procurement method and this largely depends on whether the sub-contractors make a contribution to design. If so, their procurement costs may be high because they often assist a D&B contractor bidding for a project in the design of their component. They do not charge for it on the understanding that they will obtain the work if the main contractor is successful. In other words, they contribute to the design process at their own risk, rather than being paid for design.

The supplier of bespoke components spends even more on selling in the commercial process; nearly 9% of annual turnover. This high figure is due mainly to marketing costs being about ten times higher than those of all the other participants in the process and with high sums spent also on tendering.

A contractor within a PFI-SPV spent about 5.5% of turnover on the commercial process; about double that of main contractors, almost all of it at the tendering stage. Table 22 shows that overall commercial costs for PFI are much higher than for any other form of procurement considered.

These findings demonstrate that contemporary moves towards collaborative working arrangements are characterized by the early involvement of contractors, and some specialist trade contractors, in the design process. The aim of this early involvement is two-fold: first, to enable those who are going to fabricate the buildings and its parts to make a constructive input into the design process and second, to enable clients to understand better what they are buying and how it will meet their business needs. In many cases, contractors are taking responsibility for design, even if they are not carrying it out. But by moving design responsibility into the contractor's organizations, and coupling this with an active role in the design process, contractors and their suppliers are carrying out larger amounts of pre-construction work for increasing amounts of time, often at their own risk, and with some uncertainty as to whether the project will go ahead and bear any income for them. This is what lies at the heart of their worries about the costs of tendering. While today's industry has successfully moved away from its earlier reliance on contractual disputes and litigation, this has been replaced by high up-front cost in negotiating frameworks and various types of partnering arrangement, based on the promise of future work. If the future work is forthcoming, this is not a problem. But in too many cases, these expensive arrangements have not resulted in work. Clients' aspirations may be cut back, especially in the public sector where many spending programmes have been reduced before they have even started. In other cases, the pre-contract negotiations are subject to abuse as clients cynically exploit potential design solutions from competing contractors, and use them without paying for them. These practices fly in the face of collaborative working and the demand side should not expect the supply side to continue to increase its inputs into projects at its own risk. Pre-contract has to be paid for, one way or another, and clients do not make any savings by not paying for it. The cost of abortive work is merely spread over the jobs that are won. As pre-contract involvement increases, the case for involving more than a few bidders disappears.

5.2 INDIRECT COSTS OF COLLABORATIVE VS COMPETITIVE WORKING

Over the last 40-50 years, contractors' workforces have disappeared and have been replaced by labour-only sub-contracting in particular, and sub-contracting in general. This, of itself, is a severe problem for the industry because it threatens the security or continuity of employment. In civil engineering it is a major problem. One potential spin-off from partnered supply chains, where sub-contractors are assured of continuity of work, is that the continuity of work enables everyone to invest in training and development of a skilled workforce.

Collaborative working suits very well the continuing, commercial client, but may not be suitable for the one-off client, or for the private, small client. The small-scale, private, individual client frequently buys small construction work in a particular pattern: they buy from a contractor who takes responsibility for the whole job, including the design, and who then sub-contracts much of the construction work and commits to providing everything necessary, only being paid once it works. This is in stark contrast to the method for larger and commercial projects, which involves the detailed specification of builders' work and materials. At the other end of the spectrum of complexity and size, large-scale PFI projects are specified based upon what a facility will do, what it will achieve, rather than based upon a specification of builders' work and materials. Thus, small works and large works tend to be purchased based on their output, whereas those in the middle tend to be purchased based on contractor's inputs. This contrast between input specifications and output specifications is interesting because it may appear attractive to a client to eliminate the need to specify a builder's work, resulting in cheaper documentation for the project and less of a need to understand the details of the construction process. But the costs of specification are not eliminated, merely transferred along the supply chain. And if multiple bidders are preparing detailed designs, the costs may be higher, not lower, as more people would be preparing designs and solutions with only a chance of being awarded the project.

5.2.1 Dangers of collaborative approaches

Collaborative working comes in many guises. In some cases each contractor takes a turn when new projects are launched; in others there is some form of competition among selected contractors, and, in yet others, the client chooses according to the nature of the work and contractors' commitments. There is a continuum of open competition, selective competition, negotiation, collaboration and, ultimately, perhaps collusion. A question that should be raised is "Is your collaboration really collusion?" Perhaps the answer is dictated by looking at where the power resides; in the client (collaboration) or in the suppliers (collusion). With these innovative methods, competition is being reduced and, maybe, collusion made more likely. Perhaps, while the market is not buoyant, collaboration is not as likely to lead to collusion as it might during a boom, when sellers would have more power because of the ability to pick and choose their work and their clients. Clearly, the state of the market has quite an impact on the dominant models of procurement.

5.3 MAIN FINDINGS

The most important factors that influence whether firms adopt different working practices are not the costs of tendering or of winning work. If firms choose a preferred way of working, it is not because they are economizing on the commercial costs. Indeed, hardly anyone knows how much it costs them to do business. Moreover, the variability and diversity of these costs mean that the competitive process that would sift out inefficient practices will not be systematically biased for certain modes of procurement. This is very interesting because it contradicts theories such as transaction cost economics, which suggests that the driving force is the economizing of the costs of transactions.

5.3.1 Costs to bidders

While numbers can be prepared to represent the costs of bidding, they are too diverse to be useful. Averages do not make sense when they combine public sector with private sector work, design and build with traditional, partnered relationships with open competition. In the long run, data collection may continue so that with sufficient data, such comparisons can be made. But from the work already done a number of things are clear.

The cost to a bidder of preparing a competitive price involves pitching price levels as low as possible in order to stand any chance of winning the competition. This requires fine judgement and good information, and the information is expensive to prepare. While this may only represent a fraction of a per cent of the potential contract value, if it has to take place five times to win a project, it can represent 1-2% of a contractor's turnover, more if designs have to be prepared. Remarkably, contractors have very low margins on their work, largely because most of their work is sub-contracted, but also because their profit is closely connected with their cash flows; interim payments reduce the amount they need to invest in a project. However, contractors typically plan on mark-ups of only a few per cent so the costs of competitive tendering may wipe out any returns they may have got from doing work, rendering the whole process somewhat pointless.

Thus, these costs are near to the profit margins of contractors, and the nature of contracting is such that they are more significant to construction firms than they would be to manufacturing firms, for example. Moreover, these costs can definitely be reduced. Wasteful practices that can be eliminated include:

- Anonymity of competition
- Excessively long tender lists
- Diverse pre-qualification practices
- Poor quality and late documentation provided to bidders

5.3.2 Opportunity costs for businesses

While tendering and getting work is a significant financial burden, the bulk of this work is carried out by the most senior members of the business. They spend a disproportionate amount of their time dealing with getting work. The number of estimators is a key limiting factor to contractor's capacity, not the construction workforce.

5.3.3 Direct costs to clients

Clients pay a direct cost in tendering; the preparation of designs and specifications so that the work can be put out to bid. These costs are made of fees for architects and other consultants such as quantity surveyors as well as the internal cost of having staff involved in briefing and managing the process. It must be pointed out that quantity surveyors are very different from designers. While they provide advice and information, the production of a bill of quantities is increasingly being seen by clients as an unnecessary cost. But by cutting out on these direct costs, they may find that they cause every bidder to have to produce their own documentation, and therefore the client will pay much more indirectly than would otherwise have been the case.

5.3.4 Disadvantages of collaborative working practices

Many firms spend a lot in negotiating collaborative working arrangements based on the promise of a certain volume of work, but then the spending programme turns out not to be as high as promised. Some frameworks are well known for being very expensive for getting into even though no projects flowed from them.

Significant main contractors and trade contractors have got into severe financial difficulty, including insolvency, even though (perhaps because) they had strategic partnering agreements with major clients. This raises the question of how much financial risk a contractor can carry while waiting for returns from up-front investment in their clients' projects. It also highlights the question of whether the collaboration sought is only one-way.

While the development of partnered supply chains is generally looked upon as a good thing, the way that clients restrict their options for the duration of these arrangements means that there is far less work available in competition. It can be disadvantageous to the market as a whole if the opportunities to win work are severely curtailed.

5.3.5 Preferred procurement methods and preferred selection methods

One of the surveys carried out asked respondents from across the supply chain to state their preferences for procurement methods. Clients, contractors and trade contractors all expressed a clear preference for partnering and negotiation as methods for selection and procurement of construction work. On the other hand, consultants preferred traditional methods of competitive selection and independent design – but only for contractors. Their least preferred method of procurement was design and build. However, when it came to consultants getting work, they prefer to be in long-term, partnered relationships and they would not want not be selected based on lowest price. While there are clearly good reasons for these preferences, this does raise the question of whether consultants are dependent on traditional means of procurement. With the widespread involvement of contractors in the design process, and the displacement of general contracting by design and build, there seems to be a reduction in demand for the traditional roles of consultants.

5.3.6 Public sector vs private sector

There is some evidence that public sector projects are more expensive to bid for than those in the private sector. This may be because of the need for the public sector to follow procedures and approval processes. In many public sector projects, there are huge numbers of different groups of people who need to be involved in the design and decision process. This complexity needs to be managed and there are already plenty of techniques in the management literature about how to co-ordinate numerous stakeholder groups so that they can interact with a complex project more effectively.

Another feature of the public sector is that the number of bidders seems to be higher than in the private sector. The needs for accountability seem to drive this apparent perception that a price can only be justified if it is the cheapest. There has been a lot of progress in many areas of public spending in getting away from lowest-price tendering, but there are still areas of the public sector where old attitudes and beliefs linger. For example, some public sector clients, particularly local authorities, are unwilling to talk to supply chain firms about future projects.

The Office of Government Commerce (OGC) has been instrumental in encouraging many central government departments to use collaborative approaches to construction procurement. In many areas, government clients are at the forefront of these developments and have led by example. Recently, the remit of the OGC has been expanded to cover the wider public client, including local authorities.

There is a fundamental difference between public sector and private sector clients. The purpose of the public sector is to spend public money on services for the public good. By contrast, the purpose of the private sector is to make money. These aims are not mutually exclusive, but there are reasons why the public sector should behave differently from the private sector, and therefore there should be no surprise if there continue to be differences between the two.

5.4 RECOMMENDATIONS

The conclusions lead to eight recommendations for practitioners to consider when they contemplate how to organize the construction process, and one recommendation for the academic community to reflect upon in thinking about the significance of transaction cost economics for the construction sector.

5.4.1 Early involvement of contractors/suppliers

The greatest efficiencies in the construction process lie in continuity of work and in pre-planning the construction work and the utilization of resources. Effective pre-planning requires the early involvement of the contractor, in order for the design and construction process to be effectively integrated. This, in turn, requires that selection criteria cannot be based on price for specified items of work. Alternative selection criteria need to be articulated very clearly so that competition and access to markets is not compromised. Early involvement enables those who will assemble, build and install to contribute to efficient and workable designs. Trade and specialist contractors are rarely involved in these early stages.

5.4.2 Reimburse costs for cancelled projects

Involving contractors and suppliers in design work makes the whole process more expensive for them. They don't mind, if there is a genuine chance of winning the job. Nothing rankles more than to invest a lot of effort and resources into a bid for a project that does not go ahead. It may not be a deliberate strategy on the part of the client, but clients should be open about their procurement decisions and provide some kind of financial redress in the event of project cancellation. If the market picks up to the point that supply cannot keep up with demand, contractors may be interested in investigating whether there might be a place for a bond from potential clients, payable upon cancellation of the project.

5.4.3 Selection on value rather than price

Since early involvement requires appointment before scope is agreed, selection must be on a basis other than price. For many clients of construction, this demands a move away from well-worn and familiar practices. Selection on lowest price is a decision that is easy to defend. It is objective and impartial. Selection on value sounds appealing, but the simple statement belies the complexity, particularly in the public sector, of dealing with accountability issues. Society urgently needs to consider how public servants can be given the capacity to make value judgements in a way that is accountable by some means other than cost or expense. One way might be to investigate changing the nature of accountability so that it is based on what has been achieved, rather than on how much money was spent, but this is not a trivial matter and deserves serious investigation.

5.4.4 Don't strike off contractors who cannot bid

An invitation to submit a bid may go to a contractor who is not in a position to do the job. It is widely believed among many in the supply chain that having been invited, failure to submit a bid would result in no further invitations to bid, effectively being struck off future tender lists. It is not clear why this should be so, and there are cases where contractors successfully explain to clients why they are too busy to put in a serious bid, but one response of a contractor not wishing to win a project for which it has been invited to bid is to submit a "cover price". Cover prices are obtained by submitting a price above a competitors' known price, and, as such, this is a collusive practice because the contractor submitting a cover price has to ascertain competitors' pricing levels. The motivation for doing this is that it is cheaper than calculating the real price, only to pitch it at a losing level. If contractors and other suppliers were able to explain to their customers that they may be too busy to give a realistic market price, their customers could invite someone else to submit a genuinely competitive price, but this requires an atmosphere of trust and openness that is not always present.

5.4.5 Tender only 2-3 for collaborative projects

Although it has long been accepted that the number of bidders in competitive tendering should be kept low, at around six, this was based on the assumption that designs and specifications had already been prepared and contractors were pricing bills of quantity. As fewer clients are paying for bills to be prepared, and contractors are getting more involved in the design process, the scale of their investment in potential jobs is growing. Thus, in collaborative working arrangements, contractors should be selected before the design process begins, and from much shorter lists of no more than three.

5.4.6 Standardize pre-qualification practices

The task of choosing tenderers is not easy if they are not merely pricing a pre-prepared design. If contractors are going to be seriously invited to contribute their skills, experience and judgement, then the choice of who to invite becomes very important. There is a danger that pre-qualification becomes as complex and expensive as tendering itself. Indeed, in the market for consultancy work, it is already the case that consultants spend as much on pre-qualifying for a project as they spend on competing for the project after they have pre-qualified. Worse, every client (and every public sector department or agency) has its own pre-qualification procedure. The whole industrial sector should develop more systematic and standardized approaches to pre-qualification.

5.4.7 Tell bidders who they are competing with

There is a tradition in the construction sector of keeping the identity of competing contractors from each other. This tradition had grown around the need to prevent contractors from colluding in the setting of prices. Now that contractors are preparing designs, not just prices, and being selected on a range of different criteria, it is important that they know who they are competing with. Telling bidders how many they are competing with, as well as who they are competing with, will enable them to make an informed business judgement about whether to bid and how to bid. Indeed, the idea that contractors are not supposed to know who they are bidding against is unrealistic. It takes a contractor about ten minutes to find out who else is on the tender list. Even if identity of other bidders can be concealed, it may not be a good idea. This is because responsible and professional contractors will not know if they are up against opportunistic firms without the technical and economic capacity to do the work. Contractors need to know whether they can afford to invest the resources needed to bid the job. There is a case that they should know whether they have a real chance of winning.

5.4.8 Produce timely informative documents

The documents on which contractors are invited to bid are frequently incomplete or simply missing from tender packages. This seems to come about due to various sloppy practices on the part of consultants and their clients. It was suggested that some consultants may be over-ambitious in terms of how soon they can get the project to tender stage. This may account for why it is that tender invitations are not always fully documented. If the requirements for the work are not clear, then the price can be only provisional. This can only lead to friction and mistrust later in the project, as various people's expectations fail to be met.

5.4.9 Consider professional roles when re-designing procurement processes

The changes to the roles of contractors and trade contractors must have consequences for the scale and scope of consultants' work. In developing collaborative working practices, much of the attention seems to have been directed at selecting and paying contractors and getting them involved early. There is a need to develop as clear an understanding about the impact of these collaborative practices on the roles of professional consultants.

5.4.10 Transaction cost economics does not explain the construction sector

The theories collected together under the heading of transaction cost economics emerged from the study of institutional economics, which provides very useful explanations of the institutional structure of the construction sector, indeed of any industrial sector. The development of the transaction cost model into an analytical or decision-making tool is not only inappropriate, but also unworkable. While theorists bemoan the lack of empirical evidence for transaction cost economics, it turns out that the main reason for this lack of data is the impossibility of collecting such data at the level of projects or companies. No company monitors cost data in this kind of detail, so there is no basis for a detailed cost-benefit analysis of different ways of working. Moreover, on a practical level, there are many more important issues than those dealt with by transaction cost theory. So, while the cost of transactions is an important issue, it is not the deciding issue in terms of how construction work is organized and procured.

5.4.11 Finally...

There is clearly a great deal more research to be done in developing an understanding of the way that costs are incurred in the processes of design and construction. This work has shown, once again, that the simplest questions have the most complex answers. The research has generated an enormous amount of interest and enthusiasm among practitioners and academics alike, although for different reasons. Future research should focus on re-designing the commercial processes to take advantage of the skills and expertise in all sectors of the construction industries. There will never be one right answer for all circumstances, but their procedures and methods need to be tailored to suit the circumstances of each type of client, each type of project, and be adaptable to changing economic conditions.

6 References

Anumba, C J and Evbuomwam, N F O (1997) Concurrent engineering in design-build projects. *Construction Management and Economics*, **15**(3), 271-81.

Becker, D F (1993) The cost of general conditions. *AACE transactions*, G.01, AACE International, Morgantown, WV.

Benhaim, M (1997) *Inter-firm relationships in the construction industry: towards the emergence of networks? A comparative study between France and the UK.* Unpublished DBA Thesis, Brunel University.

Bennet, J and Jayes, S (1998) *The seven pillars of partnering*. London: Thomas Telford.

Betts, M (1990) Methods and data used by large building contractors in preparing tenders. *Construction Management and Economics,* **8**(4) 399-414.

Bleeke, J and Ernst, D (1995) Is your strategic alliance really a sale? *Harvard Business Review,* **71**(1), 97-105.

Buckley, P J and Enderwick, P (1989) Manpower management. *In:* Hillebrandt, P M and Cannon, J (Eds), *The British construction firm.* London: Macmillan.

Bunn, R (1996) Innovation. The search for solutions. *Building Services Journal,* **18**(7), 27-31.

Casson, M (1994) Why are firms hierarchical? *Journal of the Economics of Business*, **1**, 47-76.

Chang, C Y and Ive, G (2001) A comparison of two ways of applying a transaction cost approach: The case of construction procurement routes. *In: Bartlett Research Papers*, No. 13, 41pp.

Chau, K W and Walker, A (1994) Institutional costs and the nature of the sub-contracting in the construction industry. *CIB W92 Procurement Systems Symposium,* Hong Kong, The Department of Surveying, Hong Kong University.

Coase, R H (1937) The nature of the firm. *Economica,* **4**, 386-405.

Coase, R H (1960) The problem of social cost. *Journal of Law and Economics*, **3**, 1-44.

Connaughton, J N (1994) Value by competition: a guide to the competitive procurement of consultancy services for construction. *CIRIA Special Publication,* No. 117. London: Construction Industry Research and Information Association.

Construction Industry Board (1997a) *Partnering in the team*. London: Thomas Telford.

Construction Industry Board (1997b) *Code of practice for the selection of sub-contractors*. London: Thomas Telford.

Construction Industry Board (1997c) *Code of practice for the selection of main contractors*. London: Thomas Telford.

Cook, A E (1990) The cost of preparing tenders for fixed price contracts. *In:* Harlow, P (Ed) *Technical Information Service,* No 120. Ascot: Chartered Institute of Building.

Dahlmann, C J (1979) The problem of externality. *Journal of Law and Economics,* **22**(1), 141-62.

Dawood, N N (1994) Developing an integrated bidding management expert system for the pre-cast concrete industry. *Building Research and Information*, **22**, 95-102.

Dawood, N N (1995) An integrated bidding management expert system for the make-to-order pre-cast industry, *Construction Management and Economics*, **13**, 115-25.

Dietrich, M (1994) *Transaction cost economics and beyond.* London: Routledge.

Dillman, D A (2000) *Mail and internet survey: the total design method.* 2nd ed. New York: John Wiley & Sons.

Dorée, A, Holmen, E and Caerteling, J (2003) Co-operation and competition in the construction industry of the Netherlands. *In:* Greenwood, D J (Ed), *19th Annual ARCOM Conference*, 3-5 September 2003, University of Brighton. Association of Researchers in Construction Management, **2**, 817-26.

Duff, R, Emsley, M, Gregory, M, Lowe, D and Masterman, J (1998) Development of a model of total building procurement costs for construction clients. *In:* Hughes, W P (Ed), *14th Annual ARCOM Conference*, University of Reading, UK, Association of Researchers in Construction and Management, **1**, 210-18.

Dutta, S and John, G (1995) Combining laboratory experiments and industry data in transaction cost analysis: the case of completion as a safeguard. *Journal of Law, Economics and Organization*, **11**, 87-111.

Eccles, R G (1981) The quasi-firm in the construction industry. *Journal of Economic Behaviour and Organization*, **5**, 335-57.

Egan, J (1998) *Rethinking construction: the report of the construction task force* (The Egan Report). Department of the Environment, Transport and the Regions, London: HMSO.

Egan, J (2002) *Accelerating change: a report by the strategic forum for construction.* London: Rethinking Construction.

Emsley, M, Lowe, D, Duff, A R, Harding, A and Hickson, A (2002) Data modelling and the application of neural network approach to prediction of total construction costs. *Construction Management and Economics,* **20**(6), 465-72.

Fawcett, S E and Magnan, G M (2001) *Achieving world-class supply chain alignment: Benefits, barriers, and bridges*, Focus Study for the center for Advanced Purchasing Studies, Tempe, AZ. Available at www.capsresearch.com.

Flanagan, R and Norman, G (1989) Pricing Policy. *In:* Hillebrandt, P M and Cannon, J (Eds), *The management of construction firms: aspects of theory.* London: Macmillan.

Gordon, C M (1994) Choosing appropriate construction contracting methods. *Journal of Construction Engineering and Management, ASCE* **120**(1), 196-210.

Gray, C and Flanagan, R (1989) *The changing role of specialist and trade contractors*. Ascot: Chartered Institute of Building.

Gray, C and Hughes, W P (2000) *Building design management*. London: Arnold.

Greenwood, D J (2001) Sub-contract procurement: are relationships changing? *Construction Management and Economics,* **19**(1), 5-7.

Grimsey, D and Graham, R (1997) PFI in the NHS (private finance initiative in the UK National Health Service). *Engineering, Construction and Architectural Management*, **4**(3), 215-31.

Gruneberg, S L and Ive, G (2000) *The economics of the modern construction firm.* London: Palgrave.

Gruneberg, S and Hughes, W P (2004) Analysing the types of procurement used in the UK: a comparison of two data sets. *Journal of Financial Management of Property and Construction,* **9**(2), 65-74.

Gruneberg, S and Hughes, W P (2005) UK construction orders and output: a comparison of government and commercially available data 1995-2001. *Journal of Financial Management of Property and Construction*, **10**(2), 83-93.

Harrison, R S (1987) Managing the estimating function. *In:* Harlow, P (Ed), *Technical Information Service*, No. 75. Ascot: Chartered Institute of Building, 1-5.

Her Majesty's Treasury (1999) *Procurement strategies. Procurement Guidance No. 5.* London: HM Treasury.

Hillebrandt, P M (1984) *Analysis of the British construction industry*. London: Macmillan.

Hillebrandt, P M (2000) *Economic theory and the construction industry*. 3 ed. London: Macmillan.

Hillebrandt, P M and Cannon, J (1990) *The modern construction firm.* London: Macmillan.

Hillebrandt, P M and Hughes, W P (2000) What are the costs of procurement and who bears them? *In:* Ngowi, A B and Ssegawa, J (Eds), *Challenges facing the construction industry in developing countries, 2nd International Conference of the CIB Task Group 29*, 15-17 November 2000, Gabarone, Bostwana. Botswana National Construction Council (BONCIC, Faculty of Engineering and Technology, University of Botswana, Council for Research and Innovation in Building (CIB), Vol. 1, 415-20.

Holmstrom, B R and Tirole, J (1989) The theory of the firm. *In:* Schmalensee, R and Willig, R D (Eds), *Handbook of industrial organization.* Vol 1. Amsterdam: Elsevier Science Publishers.

Hughes, W P, Gray, C and Murdoch, J R (1997) *Specialist trade contracting: report.* Special Report 1997. London: Construction Industry Research and Information Association.

Hughes, W P, Hillebrandt, P M and Murdoch, J R (1998) *Financial protection in the UK building industry: bonds, retentions and guarantees.* London: Spon.

Hughes, W P and Hillebrandt, P M (2003) Construction industry: historical overview and technological change. *In:* Mokyr, J (Ed.-in.chief), *The Oxford Encyclopedia of Economic History.* Oxford: Oxford University Press, Vol 1, 504-12.

Jackson-Robbins, A (1999) Selecting contractors by value. *CIRIA Special Publication 150.* London: Construction Industry Research and Information Association.

Joseph Rowntree Foundation (1999) *Modernizing local government.* no 419/Apr 1999. London: Joseph Rowntree Foundation.

Kay, N M (1982) *The evolving firm: strategy and structure in industrial organization.* London: Macmillan.

Kumaraswamy, M M and Dissanayaka, S M (1998) Linking procurement systems to project priorities. *Building Research and Information,* **26**(4), 223-38.

Lambert, D M, Cooper, M C and Pagh, J D (1998) Supply chain management: implementation issues and research opportunities. *International Journal of Logistics Management*, **9**(2), 1-19.

Latham, M (1993) *Trust and money: the interim report of the joint government/industry review of procurement and contractual arrangements in the UK construction industry.* London: Department of the Environment.

Latham, M (1994) *Constructing the team: Final report of the government/industry review of procurement and contractual arrangements in the UK construction industry.* London: HMSO.

Lemessany, J and Clapp, M A (1975) *Resource inputs to new construction: the labour requirements of hospital building* CP 85/75. Watford: Building Research Establishment.

Levene, P, Jackson, N , Gray, R , Jensen, J , Massey, G , Moschini, S , West, R and Woodman, R (1995) *Construction procurement by government: An efficiency unit scrutiny (the Levene report)*. London: Efficiency Unit Cabinet Office.

Lingard, H, Hughes, W P and Chinyio, E (1998) The impact of contractor selection method on transaction costs: a review. *Journal of Construction Procurement*, **4**, 89-102.

Ljungberg, A (1998) *Measurement systems and process orientation with focus on the order process*. Department of Engineering Logistics, Lund University.

Lyons, B R (1994) Contracts and specific investment: an empirical test of transaction cost theory. *Journal of Economics and Management Strategy,* **3**, 257-78.

Lyons, B R (1995) Specific investment, economies of scale, and the make-or-buy decision: a test of transaction cost theory. *Journal of Economic Behaviour and Organization*, **26**, 431-43.

Masden, S E, Meehan, J W and Snyder, E A (1991) The costs of organization. *Journal of Law, Economics and Organization*, **7**(1), 1-25.

Matthews, J, Tyler, A and Thorpe, A (1996) Pre-construction project partnering: Developing the process. *Engineering, Construction and Architectural Management*, **3**(1/2), 117-31.

Miller, C J M, Packham, G A and Williams, T (1999) Re-defining sub-contracting: reducing transaction costs? *In:* Hughes, W P (Ed), *15th Annual ARCOM Conference*, September 15-17, Liverpool John Moores University, UK. Association of Researchers in Construction and Management, **2**, 655-64.

Ministry of Works (1944) *The placing and management of building contracts.* Report of the Central Council for Works and Buildings to the Minister of Works (Chairman Sir Ernest Simon). London: HMSO.

Moore, D R and Dainty, T (2001) Intra-team boundaries as inhibitors of performance improvement in UK design and build projects: a call for change. *Construction Management and Economics,* **19**(6), 559-62.

Murray, M and Langford, D (Eds) (2003) *Construction reports 1944-98.* Oxford: Blackwell.

National Joint Consultative Council (1994) *Codes of Procedure for Tendering* (various). London: NJCC.

Nettleton, B (2000) Best value and direct services. *ICE Proceedings: Municipal Engineer*, **139**(2), 83-90.

Ng, S T, Kumaraswamy, M M and Chow, L K (2001) Selecting consultants through combined technical and fee assessment: A Hong Kong study. *In:* Akintoye, A (Ed), *17th Annual ARCOM Conference*, 5-7 September 2001, University of Salford. Association of Researchers in Construction Management, **1**, 639-48.

Pasquire, C L and Collins, S (1996) The effect of competitive tendering on value in construction. *RICS Research Papers*, **2**(5), 1-32.

Pass C, Bryan, L , Pendleton, A and Chadwick, L (1995) *Collins dictionary of business*. 2 ed. Glasgow: Harper Collins.

Pearson, G (1985) Tender assessment. *Chartered Quantity Surveyor*. **8**, 194-5.

Pitelis, C (1993) *Transaction costs, markets and hierarchies*. Oxford: Blackwell.

Poh, P S H and Horner, R M W (1995) Cost-significant modelling: potential for use in South-East Asia. *Engineering, Construction and Architectural Management,* **2**(2), 121-39.

Pokora, J and Hastings, C (1995) Building Partnership: a team working and alliances in the construction industry. *In:* Harlow, P (Ed), *Construction Papers No 5.* Ascot: Chartered Institute of Building.

Rahman, M M and Kumaraswamy, M M (2002) Joint risk management through transactionally efficient relational contracting. *Construction Management and Economics,* **20**(1), 45-54.

Reve, T (1990) The firm as a nexus of internal and external contracts. *In:* Aoki, M, Gustafsson, B and Williamson, O E (Eds), *The firm as a nexus of treaties*. London: Sage.

Reve, T and Levitt, R E (1984) Organization and governance in construction. *International Journal of Project Management*, **2**, 17-25.

Roberts, J S (2003) The buzz about supply chain. *Inside Supply Management*, **14**(7), 24.

Shen, L and Song, W (1998) Competitive tendering practice in Chinese construction. *Journal of Construction Engineering and Management, ASCE* **124**(2), 155-61.

Suraya, I (1997) *A feasibility study in quantifying transaction costs in construction procurement routes in the UK: the case of general contracting and integrated design-and-build.* Unpublished MSc Thesis, Department of Construction Economics and Management, Bartlett School, University College London.

Svensson, R (2001) Success determinants when tendering for international consulting projects. *International Journal of the Economics of Business*, **8**(1), 101-22.

The Consultancy Company Ltd. (1997) *Evaluation of the Construction (Design and Management) Regulations 1994*. HSE Contract Research Report 158/1997. Sudbury: HSE Books.

The HCi Journal, (2003) Customer/Supplier Map *http://www.hci.com.au/hcisite2/toolkit/customer.htm* Accessed 25 August 2004

Thompson, J D (1967) *Organizations in action: Social science bases of administrative theory*. New York: McGraw-Hill.

Tombesi, P (1997) *Travels from flatland*. Unpublished PhD Thesis, University of California, Los Angeles, USA.

Turner, J R and Simister, S J (2001) Project contract management and a theory of organization. *In: Erasmus Research Institute of Management Report Series* (ERS-2001-43-ORG). Rotterdam: ERIM.

Uher, T E (1996) Cost estimating practices in Australian construction. *Engineering, Construction and Architectural Management*, **3**(1/2), 83-95.

Wåhlström, O (1991) Simplified tender documents, giving an unambigous representation of the finished building. *Building Research and Information*, **19**(5), 311-14.

Wang, M T and Wu, T S (2000) Cyberspace tendering system, and electronic procurement issues. *International Journal of Computer Integrated Design and Construction,* **2**(2), 134-41.

Wang, S Q, Tiong, R L K , Ting, S K , Chew, D and Ashley, D (1998) Evaluation and competitive tendering of BOT power plant project in China. *ASCE Journal of Construction Engineering and Management*, **124**(4), 333-41.

Williamson, O E (1975) *Markets and hierarchies: analysis and antitrust implications: a study in the economics of internal organization*. New York: The Free Press.

Williamson, O E (1979) Transaction cost economics: the governance of contractual relations. *Journal of Law and Economics,* **22**, 233-61.

Williamson, O E (1985) *The economic institutions of capitalism*. London: The Free Press.

Williamson, O E (1990) The firm as a nexus of treaties: an introduction. *In:* Aoki, M, Gustafsson, B and Williamson, O E (Eds), *The firm as a nexus of treaties*. London: Sage.

Winch, G (1989) The construction firm and the construction project: a transaction cost approach. *Construction Management and Economics*, **7**, 331-45.

Winch, G (1995) *Project management in construction: towards a transaction cost approach*. Le Groupe Bagnolet, Working Paper 1. University College London.

Wong, C H, Holt, G D and Cooper, P A (2000) Lowest price or value? Investigation of UK construction clients' tender selection process. *Construction Management and Economics*, **18**(7), 767-74.

Zarkada-Fraser, A (2000) A classification of factors influencing participating in collusive tendering agreements. *Journal of Business Ethics*, **23**(3), 269-82.

7 Indexes

7.1 NAME INDEX

7.2 SUBJECT INDEX

Appendix A: Glossary of terms

bid: contractor's tender (q.v.)
bid peddling: generally restricted to sub-contract procurement, involves the successful main contractor receiving unsolicited sub-contract bids directly after securing a project.
bounded rationality: construct proposed by Williamson (1975) as a human factor that accounts for transaction costs. Situation where knowledge is limited by lack of information that would be either costly, or in some cases impossible to acquire at the time. This is particularly true of the long, complex, contingent exchange transactions that typify modern market activity.
competitive tendering: method of competition where the criterion for selection is predominantly (though not exclusively) the price tendered.
complexity: a construct proposed by Williamson (1975) as an environmental factor accounting for transaction costs. Organizational complexity increases to cope with environmental uncertainty (q.v.) (see Thompson 1967).
Dutch auctioning: the practice whereby bidders are invited to engage in further cost-cutting rounds after the initial tenders have been received.
estimate: a reasonably accurate calculation of the probable cost of carrying out work. For a prudent contractor, calculation of an estimate is usually a pre-requisite to tendering.
governance structure: any one of a number of possible organizational forms.
hierarchies: the position at one extreme of the continuum of possible organizational forms (opposite to markets q.v.) where all exchange is conducted within a single integrated organization.
imperfect market: construct proposed by Williamson (1975) to be an environmental factor accounting for transaction costs. Although markets are assumed to be perfect in classical economic theory, neo-classical economics recognizes that small numbers situations can exist to interfere with the expected working of a perfect market.
information asymmetry: construct proposed by Williamson (1975) as a transaction-specific factor that accounts for transaction costs. In classical economics individuals who transact are assumed to have perfect access to information; in reality, their access to information is potentially unequal.
markets: the position at one extreme of the continuum of possible organizational forms (opposite to hierarchy q.v.) where exchange is conducted between a large number of independent firms.
mark-up or margin: a part of the tender sum that represents an addition to the estimate, usually in the form of a percentage, to recover the cost of business overheads and the required profit return.

negotiated tender: a method of allocating work whereby the agreed project price is arrived at by means other than price competition. This can take the form of price negotiations with a single contractor, or parallel negotiations with several contractors.

network: an intermediate relationship lying on the continuum of possible forms lying between markets and hierarchies (see Williamson 1985).

opportunistic behaviour: construct proposed by Williamson (1975) as a human factor that accounts for transaction costs. Opportunism is the underlying human tendency to take advantage of situational factors that limit the workings of the market, such as imperfect markets (q.v.) or information asymmetry (q.v.).

open tendering: a form of competitive tendering (q.v.) where competition is unrestricted, and the client is open to any bona fide offer to carry out the work.

overheads (business overheads; head-office overheads): the fixed costs of a contractor being in business.

partnering: a method of transforming what would normally be a contractual relationship into an obligational one. Partnering can be project-specific (where the transformation takes place post-tender) or strategic (where the allocation of work is itself by non-traditional means, e.g. by negotiation).

pre-qualification: stage in the tendering process in which contractors undergo qualitative selection in order to enable the client to produce a tender short-list.

pre-tender costs: the costs (to both client and contractors) of arriving at tender stage. For the client, this includes the costs of preparing tender documents, obtaining information and conducting pre-qualification processes and inviting tenders. For the contractor, this includes the costs of marketing and pre-qualification.

post-tender costs: the costs (to both client and contractors) of performing the contract following acceptance of the tender. These include the costs of post-tender negotiations, contract planning / drafting, performance monitoring, performance enforcement and prosecuting disputes.

quasifirm: term given by Eccles (1981) to an organizational relationship intermediate between market and hierarchy.

selective (selective list) tendering: a form of competitive tendering (q.v.) where the client chooses to restrict competition to a selected number of contractors (usually between four and six). This commonly involves a pre-qualification (q.v.) procedure for those wishing to make the tender list.

special purpose vehicle (SPV): a firm set up with the sole purpose of producing and operating a facility to provide a service over a specific period. Usually a joint venture between various combinations of funders, contractors, facilities managers and operators.

tender: An offer to carry out work for a stated price. Thus, for the client, tendering traditionally involves both selecting a contractor and agreeing a price.

tendering costs: the costs (to both client and contractors) of preparing for and submitting tenders. For the client, this includes the costs of administering the tender process. For the contractor, this includes the costs of calculating prices, assessing risks and formulating environmental, health and safety, and quality plans.

transaction(s) cost analysis (theory): theory from institutional economics first proposed by Coase (1937, 1960) and refined by Williamson (1975, 1979, 1985, 1990). Attempts to explain why firms (and hence contractual exchange) exist at all.
transaction(s) costs: the various costs, proposed by Coase (1960: 22) of market transactions. These include the costs of negotiations leading up to the bargain; of drawing up the contract; of undertaking the necessary inspection and supervision to enforce compliance; and of conducting disputes.
turnover: The value of a contractor's annual sales.
two-stage tendering: a form of contractor selection involving two stages: one based on price competition, and the other on negotiation. The two stages may take place in any order, depending on the requirements of the client.
uncertainty: construct proposed by Williamson (1975) as an environmental factor accounting for transaction costs. Uncertainty has long been recognized as a key contingency in an organization's response to its environment (see Thompson 1967).

Appendix B: Annotated bibliography

Akintoye, A (1994) Design and build: a survey of construction contractors' views. *Construction Management and Economics*, **12**(1), 155-63.
Abstract: Design and build (D & B) has become a popular mode of procuring construction work. A total of 52 construction firms responsible for 25% of UK construction output for 1991 were surveyed using a structured questionnaire to investigate their current views on this procurement route. Novated D&B is widely used although not favoured by contractors. The contractors would like consultants to continue to provide them with concept design and specification and would rather support the *develop and construct* technique. The usage of D&B on private sector projects is around 21% of workload from this sector compared to12% in the public sector.
Notes: It should be noted that figures showing the relative take up of D&B in different sectors were taken from a 1991 survey of 52 firms.

—— **(1994)** Design and build procurement method in the UK construction industry. *In:* Rowlinson, S (Ed), *CIB W92 Procurement Systems Symposium*, 4-7 December 1994, Hong Kong. The Department of Surveying, Hong Kong University, Vol. 1, 1-10.
Abstract: This paper reports the results of a recent comparative analysis of the opinions of both the UK architects and contractors on design and build procurement methods for building projects. Design and build is perceived by architects as producing poor quality building. This is not the view of contractors. A majority of architects and contractors are of the opinion that design and build offers time savings in the order of 5 to 20%. Novated D&B and Develop & Construct are increasingly popular variants of Design & Build, in preference to the conventional (pure) D&B because they offer clients more control over design process and quality. Design and build contracts account for 8% and 12% of architects' and contractors' public sector workload respectively and 19% and 20% of their private sector workload respectively.
Notes: As above, it should be noted that figures showing the relative take-up of D&B in different sectors were taken from a 1991 survey.

—— **(2000)** Analysis of factors influencing project cost estimating practice. *Construction Management and Economics*, **18**(1), 77-89.
Abstract: Although extensive research has been undertaken on factors influencing the decision to tender and mark-up and tender price determination for construction projects, very little of this research contains information appropriate to the factors involved in costing construction projects. The object of this study was to gain an understanding of the factors influencing contractors' cost estimating practice. This was achieved through a comparative study of 84 UK contractors classified into four categories, namely, very small, small, medium and large firms. The initial analysis of the 24 factors considered in the study shows that the main factors relevant to cost estimating practice are complexity of the project, scale and scope of construction, market conditions, method of construction, site constraints, client's financial position, buildability and location of the project. Analysis of variance,

which tests the null hypothesis that the opinions of the four categories of companies are not significantly different, shows that except for the procurement route and contractual arrangement factor there is no difference of opinion, at the 5% significance level, on the factors influencing cost estimating. Further analysis, based on a factor analysis technique, shows that the variables could be grouped into seven factors; the most important factor grouping being project complexity followed by technological requirements, project information, project team requirement, contract requirement, project duration and, finally, market requirement.
Notes: This confirms the complexity of the estimating process and also the element of judgement and appraisal of difficult-to-measure factors.

Akintoye, A, Beck, M, Hardcastle, C, Chinyio, E A and Assenova, D (2001) The financial structure of Private Finance Initiative projects. *In:* Akintoye, A (Ed), *17th Annual ARCOM Conference*, 5-7 September 2001, University of Salford, UK. Association of Researchers in Construction Management, Vol. 1, 361-9.
Abstract: Over the past years PFI has become a key approach to public-private partnership in the UK. Unlike other forms of co-operation between the public and the private sector, PFI involves the investment of private capital in projects, which provide services to the public. This paper focuses on the financial structure of the PFI schemes and explores the financing options available to PFI participants. Based on 48 interviews with senior representatives from leading UK international companies, the authors identify the currently predominant form of finance through senior and subordinated debt, which they contrast with the use of bond finance and some rarely used financing methods like lease finance and Mezzanine finance. They note that the convergence of PFI financial practices allows for the use of financial models, which lend themselves to standardization. The paper concludes with the proposition that a greater standardization of the planning practices associated with PFI could help counter criticisms of PFI which view this type of procurement as inefficient and costly.
Notes: The (current) lack of standardized approaches to PFI bids results in increased costs of tendering.

—— (2001) *Framework for risk assessment and management of Private Finance Initiative* 1 903661 28 5, Glasgow: Glasgow Caledonian University.
Abstract: The study of PFI participants and practices started in April 2000 and lasted until August 2001. Initially, PFI participants were observed on a non-project-specific basis. However, from January 2001, the research translated into project-specific case studies. As observations were made during the two phases, substantial findings emerged. Ninety-four interviews were held with diverse PFI participants that included contractors, financial firms, facilities management organizations, independent consultants and public sector clients. The interviews ascertained how they identified, evaluated, reported and managed the risks of their PFI projects.
Notes: The study observed, *inter alia*, that due to high bidding costs, the detailed assessment of risks is delayed until a consortium has become the preferred bidder.

Akintoye, A, Bowen, P A and Hardcastle, C (1998) Macro-economic leading indicators of construction contract prices. *Construction Management and Economics*, **16**(2), 159-75.
Abstract: An understanding of future trends in construction prices is likely to influence the construction investment strategy of a variety of interested parties, ranging from private and public clients to construction contractors, property speculators, financial institutions, and

construction professionals. This paper derives leading indicators for construction prices in the United Kingdom. These indicators are based on two experimental methods: turning points of the basic indicators in relation to construction price turning points; and predictive power of lags of the basic indicators. It is concluded, based on the analyses, that unemployment level, construction output, industrial production, and ratio of price to cost indices in manufacturing are consistent leading indicators of construction prices. Building cost index and gross national product constitute coincident indicators. "Popular" macro-economic time series such as nominal interest rate, inflation rate, real interest rate, all share index and money supply produced inconclusive results.
Notes: Seeks indicators of likely future trends in construction prices. No relevance to the cost of procurement other than, perhaps, the identification of one more activity that might precede the construction process.

Akintoye, A and Fitzgerald, E (2000) A survey of current cost estimating practices in the UK. *Construction Management and Economics*, **18**(2), 161-72.
Abstract: The results are documented of an investigation into current cost estimating practices of contractors for construction projects. A questionnaire survey of contractors was undertaken in which the respondents are classified into four groups based on their turnover, namely: very small, small, medium and large firms. The survey indicates that contractors, irrespective of size, continue to undertake cost estimating predominantly for construction planning purposes, including the preparation of tenders and cost control of projects during the execution stage and, to a lesser extent, for construction project evaluation. Recent developments in cost estimating methods and tools that consider risks and variability in cost estimates, such as the use of range estimating and parametric estimating techniques, have not been adopted by contractors. The practice of cost estimating does not differ from conventional techniques based on the use of labour and material constants to obtain prices for bills of quantities items on an item by item basis.
Notes: A study of cost estimating undertaken by a survey of 84 co-operating firms. It is relevant that many different persons are involved in the estimating process. Large firms have estimating departments but in addition a number of other job roles could be involved, including: managing director, contracts managers, quantity surveyors, site management, store managers/buyers, planning or programming engineers, commercial managers, design engineers, suppliers, cost planners and insurance advisors.

Akintoye, A, McIntosh, G and Fitzgerald, E (2000) A survey of supply chain collaboration and management in the UK construction industry. *European Journal of Purchasing and Supply Management*, **3-4**(6), 159-68.
Abstract: The paper details the results of a questionnaire survey of supply chain collaboration and management in the top UK construction industry contractors. The results indicate the formation of a significant number of partnerships/collaborative agreements between contractors, suppliers and clients following the publication of the Latham (1994) and Egan (1998) reports. It appears that construction supply chain management (SCM) is still at its infancy but some awareness of the philosophy is evident. Contractors identified improved production planning and purchasing as key targets for the application of SCM in construction. Barriers to success included: workplace culture, lack of senior management commitment, inappropriate support structures and lack of knowledge of SCM philosophy. Training and education at all levels in the industry are necessary to overcome these barriers.
Notes: Arguments are outlined for integrative procurement approaches.

Anumba, C J and Evbuomwam, N F O (1997) Concurrent engineering in design-build projects. *Construction Management and Economics*, **15**(3), 271-81.
Abstract: The design and build procurement route has witnessed significant growth in the UK construction industry over the last ten years. It is now being used for both private and public sector projects of varying complexity. There are several advantages associated with this method of construction procurement including shortening of lead times, involvement of the contractor in the design process, greater price certainty, improved communication and reduced construction time amongst others. Conversely, there are also a number of disadvantages ascribed to the design and build method of procurement. Some of these include reduced design quality, inhibition of changes by clients, and high tendering costs. A new process model is proposed to address many of the procurement route's present shortcomings. In particular, the model facilitates concurrent project development in the design and build process through the integration of all project participants into a multi-functional matrix team capable of resolving potential "downstream" problems early in the project life-cycle, and the provision of a formal mechanism for the improved abstraction of client requirements based on design function deployment (DFD) – a concurrent engineering design system.
Notes: Usefully distinguishes the variants of design-build and accepts predictions that its use will increase. Among other things, highlights the high costs of tendering and mentions that there have been calls for clients to pay tendering costs to unsuccessful bidders. Proposes a radical review of the existing process, without reference to the variants already identified. Concurrent engineering is proposed as a promising solution, but seems to be nothing more than taking account of all aspects of use, throughout the life cycle, during early design. A new model is developed, accompanied by a variety of untested claims about its benefits. There is one mention of the impact on tendering costs. Based upon no data at all, the authors suggest that the proposed changes to the tendering procedure may be resisted, they suggest that the changes will not affect the cost of tendering, then contradict themselves in claiming that the abortive work will be less (sic). There are no data in this paper and no suggestion as to how the cost of tendering might be quantified.

Aqua Group (1990) *Tenders and contracts for building*. Oxford: Blackwell Science.
Abstract: This handbook examines the tendering procedures and contractual arrangements available to clients, and discusses the differing circumstances dictating the choice of both tendering procedures and contractual arrangements and discusses their advantages and disadvantages.
Notes: A readable handbook of tendering and estimating rather than an academic work. No mention of the comparative costs of tendering.

Ashworth, A (1996) *Pre-contract studies: development economics, tendering and estimating. Education Framework, CIOB*, Harlow, Essex: Longman.
Abstract: A student textbook. Topics discussed include: the development process; property investment economics; capital investment; financial data evaluation; development appraisal; urban land economics; economics of property investment; sources of finance; budgeting, costing and cash flows; procurement; tendering and estimating.
Notes: Chapter 10 discusses alternative procurement strategies and their advantages and disadvantages. There is little mention of the comparative costs of tendering under the different procurement strategies.

Baccarini, D (1998) Cost contingency: a review. *AIQS Refereed Journal*, **2**, 8-15.
Abstract: This paper focuses on the cost contingency in projects, presenting the attributes of contingencies, the elements that are covered by cost contingencies, and contingencies estimation, both deterministic and probabilistic, and the methods employed. It concludes that a review of the literature confirms that there is no standard definition for contingency. It is recommended that a project policy should exist with regards to contingency definition which is understood and accepted by all project participants. Finally, the paper suggests the probabilistic method of cost estimation contingency as a more robust approach and that cost contingency is a risk management tool that has to be proactively managed once it has been calculated and formally authorized.
Notes: Nothing on the actual costs of tendering.

Bajaj, D, Oluwoye, J O and Lenard, D (1997) An analysis of contractors' approaches to risk identification in New South Wales, Australia. *Construction Management and Economics*, **15**(4), 363-9.
Abstract: The process of risk identification at the tendering and estimating stage is the first stage of the risk management process, and for the risk management process to be of benefit and for the project objectives to be achieved, this stage should be very detailed and thorough. The aim of this study is to identify, investigate and evaluate the process of risk identification at the tendering and estimating stage for construction contractors in the NSW region. The data for this were collected during the months of December 1994 and January 1995 using a sample survey of a cross-section of 19 construction contractors, and the results were analysed using frequency distribution. The results show that the most frequently used methods of risk identification are the top-down approach techniques, where the project is analysed from an overall point of view. Techniques based on top-down approach lead to guesswork in terms of contingency for risks accepted by the construction contractors. Bottom-up risk identification techniques are not popular except for a questionnaire and checklist approach. Also, it was unlikely that the contractors would discuss risk allocation with the clients. All the contractors interviewed agreed that when a risk identification process is followed it improves the accuracy of their estimates.
Notes: A small opinion survey of 19 contractors in NSW, Australia, about their approaches to risk identification during the tendering and estimating processes. There was no attempt to assess the resources expended in risk identification. The purpose was to ascertain the variety of approaches taken.

Baker, S T (1990) Partnering: contracting for the future. *Cost Engineering*, **32**(4), April, 7-12.
Abstract: The benefits of partnering between an owner and a contractor are discussed, outlining the principles of partnering and its application to the aspect of contracting. Reasons why owners and contractors might want to initiate a partnering arrangement are identified. It further suggests conditions conducive to partnering from both the owner's and contractor's perspective. It concludes that partnering is not a quick-fix remedy, but its very foundation necessitates trust, good planning, patience and persistence.
Notes: No mention of comparative costs of partnering against other procurement strategies.

Barker, J (1998) Costs of tendering. *The Building Economist,* (December), 13-9.
Abstract: The author asks is there a justification for clients to pay the costs of tendering, or is this simply an overhead? This paper explores the benefits and disadvantages of payment of tendering fees. It outlines suggestions which it hopes when implemented would reduce

the cost of tendering and/or ensure clients pay a reasonable share as well as improving the tender market in general. It concludes that tendering costs are a fact and that honesty and fair dealing, along with improving tendering practices and procedures, will do more than anything else to reduce tendering costs and ensure that clients pay their fair share.
Notes: No mention of actual or comparative costs of tendering.

Becker, D F (1993) The cost of general conditions. *AACE transactions, G.01,* AACE International, Morgantown, WV.
Abstract: The impacts of certain clauses in general conditions of contract are considered. The intent is to make drafters and users of general conditions of contract aware of how certain clauses may have a cost associated with their inclusion so that they can make intelligent business decisions about whether to include them in the document.
Notes: In considering the cost associated with the general conditions of contract, this author seems to be addressing the impact that clauses and terms can have on the contract price. However, the paper simply states conclusions with no evidence, argument or support whatsoever. Thus, there are many interesting, but totally unsupported statements such as: As the risk is shifted from the owner to the contractor, the contractor will increase its indirect cost, contingencies and profit margins to cover the unknown conditions. Reliance on the wrong party for direction can be costly for both the contractor and the owner. Failure of the owner to perform its responsibilities can be costly for the owner and so on. In some cases, examples of clauses are given, and their possible impact suggested. This is the justification for the assertions about how contract conditions impact the cost of construction. Clearly, the article is based on the author's experience, but the lack of supporting evidence provides only hypotheses about what is generally applicable, rather than conclusions. There is no attempt to quantify the scale of the cost associated with the chosen contractual terms.

Benhaim, M (1997) *Inter-firm relationships in the construction industry: towards the emergence of networks? A comparative study between France and the UK*, Unpublished DBA thesis, Brunel University.
Notes: Reports on the construction industries of the UK and France. The industry structure is accounted for by the particular business strategies of main contractors. The possibilities for, and barriers to, partnering in the UK and France are described.

Benhaim, M and Birchall, D (1998) Inter-firm relationships within the construction industry: a comparative study between France and the UK. *In:* Hughes, W P (Ed), *14th Annual ARCOM Conference*, Reading University. Association of Researchers in Construction Management, 407-16.
Notes: Builds on the doctoral thesis of one of the authors (see above) and examines the possibilities for, and barriers to, partnering in the UK and France.

Betts, M (1990) Methods and data used by large building contractors in preparing tenders. *Construction Management and Economics*, **8**(4), 399-414.
Abstract: The process of submitting lump sum competitive tenders continues to be commonly practised by UK building contractors. Most tenders are based on a detailed analysis of project details and a detailed costing of parts of the work to be done. Considerable resources are being devoted to the preparation of tenders in this way. Any means of improving the efficiency of this process would be very welcome to contractors and

to the construction industry as a whole. This paper presents a documentation of methods of tender preparation in the form of a model of the tasks executed. This model is a description of the process in its most complex possible form as currently executed and does not attempt to portray the tendering process as it is typically performed. Variations within the model have been found to occur between individual contractors and for alternative means of procurement as well as for differences in project complexity. However, the model is generally representative of the means by which tenders are prepared by large building contractors in the UK.
Notes: Thorough assessment of the data that can be necessary for bidding large projects, but no evaluation of the costs of acquiring it.

Bingham, T (1998) Improper suggestions. *Building*, **263**(19), May 8, 39.
Abstract: Short article on legal topics: the European Court and Commission found traditional tendering practices in the Netherlands to be illegal, but is it necessary to harmonize all the rules in European countries.
Notes: A practice common in the Netherlands has been found by the European Court to be illegal. This article is based upon a book by Joseph Dalby, EU Law for the Construction Industry. In 1952, the 4,000 Dutch contractors organized themselves into 28 associations, collectively governed by an overall group called SPO. The SPO introduced Uniform Price Rules (UPR), which involved controlling the tendering procedure. Member associations within the regions chose a preferred bidder for each particular project. Contractors prepared their bids in the usual way, then shared their information. The bids were then adjusted so that the preferred bidder would win the competition and the price was increased by an amount that enabled the preferred bidder to refund the losers' tendering costs and make a payment to the trade association. The European Commission investigated this practice and found it contravened the procurement rules, fining the trade associations 22.5m ecu (£14.9m). The trade associations then took the Commission to the European Court, which found that the associations were infringing article 85(1) of the EC Treaty.

—— (2000) Undercutting is over. *Building*, **265**(17), April 28, 45.
Abstract: The new best-value regime promises to put creativity and innovation back into the local authority tendering process, but, as commentators point out, it may come as a shock to the established system.
Notes: Explains how the introduction and expansion of compulsory competitive tendering in local authorities stifled creativity and originality by forcing LAs to always accept lowest bid and ignore non-conforming bid. Introduces the Local Government Act 1999 as something that helps LAs to secure continuous improvement. The aim is to enable LAs to choose a "best value" solution by weighing up economy, efficiency and effectiveness of proposals, thus not channelling them into a particular method of tendering or procurement.

Bremer, W and Kok, K (2000) The Dutch construction industry: a combination of competition and corporatism. *Building Research and Information*, **28**(2), 98-108.
Abstract: The development of the contracting system in the Dutch construction industry is presented in terms of the "poldermodel" of corporatist institutions. Although stiff competition exists between construction firms, corporatist arrangements have developed in the Netherlands to reduce high costs and risks to individual firms. The implications and benefits of these corporatist arrangements for research and development, employment, training and housing investment are assessed and concluded to provide well-balanced

market competition and regulation while maintaining long-term needs. The relationships between the many parties in the construction process (clients, investors, contractors, suppliers, designers and regulatory authorities) are analysed. A particular area in which corporatist arrangements have predominated is in tendering for public works, where there is a tradition of price-rigging on the grounds that this reduces transaction costs and favours social solidarity. However, the European Commission has banned such arrangements on competition grounds. The Dutch construction industry's attempts to reform tendering arrangements on a more competitive basis, while retaining the valued aspects of the poldermodel, are assessed.
Notes: "Poldermodel" is a Dutch term indicating a political consensus in business, particularly between state and unions. One of the purported reasons for the co-operative alliances between contractors in the Netherlands was to reduce bidding costs. The practice was investigated and exposed in 2003, causing considerable anxiety and turbulence in the Dutch construction sector.

Brewer, G (1999) Tendering and award of public works contracts. *Contract Journal*, **401**(6251), November 10, 28.
Abstract: The case of Harmon CFEM Facades (UK) Limited versus the Corporate Officer of the House of Commons. The issue: the operation of the Public Works Contract Regulations 1991. The implications: where procurement is to be based on the most economically advantageous tender, clear criteria have to be set out in the contract notice, and objective scoring systems based on these criteria have to be strictly observed in determining the successful tenderer.
Notes: The broad question that such a case poses is "What costs do EC tendering regulations impose on bidders?" The sum of damages awarded to Harmon would notionally have included their tendering costs, but the court's view of Harmon's losses from being excluded from the contract were based upon loss of potential earnings from the project.

Bröchner, J (1994) Precontractual investigation and risk aversion. *Engineering, Construction and Architectural Management*, **1**(2), 91-101.
Abstract: This paper presents an analytical framework for determining efficient levels and durations of precontractual investigation. Economic efficiency in the allocation of investigation tasks between client and tenderer is shown to depend on how closely related the technologies of investigation and construction are. Moreover, risk aversion and the interest rate affect the efficient allocation. The framework is also used as a basis for an investment analysis of the balance between client's investigation efforts and expected claims in the future. Finally, the framework is used to show how the optimal length of the investigation period can be derived from the expected cash flow associated with a project over its total life cycle, from inception to demolition. Results indicate the economic potential of tailoring risk-sharing in construction procurement, according to the type of construction project and the attitudes to risk between clients and contractors.
Notes: Useful background information on one aspect of pre-contract activity, but no direct information on the costs of carrying this out.

Brook, M (1993) *Estimating and tendering for construction work*. Oxford: Butterworth-Heinemann.
Abstract: The task of the estimator is explained in detail at every key stage from early cost studies, through the preparation of the estimate, to the creation of budgets for successful

tenders. Each step is illustrated with examples and notes, and appropriate technical documentation. Over recent years there have been significant developments in construction management, notably new procurement methods, greater emphasis on innovation and partnering, a greater reliance on cost planning as a methodology, and new developments in both industry and governmental reports and guidance. This textbook addresses each of these developments in turn.
Notes: A tendering and estimating teaching text rather than an academic work.

Bunn, R (1996) Innovation. The search for solutions. *Building Services Journal*, **18**(7), 27-31.
Abstract: Future of building services research – review of BSRIA report entitled "The building services industry: A strategy for research and innovation". Main research and innovation objectives include understanding clients' needs, procurement, standards and regulations, tendering costs, design procedures, feedback and life-cycle analysis, standardization, technology and performance, health and safety, environment, energy efficiency, water usage, indoor environment, motivation, education and recruitment.
Notes: This article (1996) is a review of BSRIA's report on the future of building research. On page 29, under the heading of tendering costs, it says that "industry must discover the real cost of tendering for different types of projects, both to the individual firm and to the industry at large and investigate ways of reducing it". It also quotes tendering costs as around 1-2% of the contract value and an average of five tenders per job.

Carr, R I (1983) Impact of number of bidders on competition. *Journal of Construction Engineering and Management, ASCE*, **109**(1), 61-73.
Abstract: A contractor seeking the highest expected value from each competitive bid will lower the mark-up as the number of competitors increases. The contractor's competitors can be expected to do likewise, and their adjustments affect the contractor's expected value. Thus, the number of competitors affects a contractor's profit two-fold. The sheer number of competitors dilutes the probability of winning, and competitors' adjustment of their mark-ups undercut a contractor's own mark-up. Competitors' estimated adjustments are easily included in a bidding analysis based on a modified general bidding model or on Gates's model. Friedman's model is not compatible with adjustments by competitors. There is a natural equilibrium among competitors based on their relative costs. A contractor with lower costs will bid a higher mark-up but still a lower bid than competitors. But it is not to the competitor's advantage to undercut the opposition further. Coupled with the relative insensitivity of expected value to mark-up, this leads to stability in the competitive bidding process. The interaction within a market, the estimating of competitors' adjustments, and the impact of the number of bidders are demonstrated for the different models.
Notes: A study in the area of bidding theory initiated by Friedman. Not concerned with the actual costs of tendering.

Chan, A P C (2000) Evaluation of enhanced design and build system: a case study of a hospital project. *Construction Management and Economics*, **18**(7), 863-71.
Abstract: This study involves an investigation of a hospital project which adopted an "enhanced design and build" form of building procurement. The paper assesses the procurement system from the perspectives of the client, client's consultants, contractor, contractor's consultants, and contractor's sub-contractors. A detailed case study of North District Hospital is described to illustrate the process of this procurement system. All the interviewees generally agreed that the hospital project was successful in meeting the time,

cost, quality, functional and safety requirements set by the client. As the first project adopting the "enhanced design-build" procurement system in Hong Kong, the benefits of applying this innovative procurement system were demonstrated.
Notes: Illustrates a variant of D&B procurement. Not concerned with the costs of tendering or the costs of procurement as such.

Chan, A P C and Tam, C M (1994) Design and build through novation. *In:* Rowlinson, S (Ed), *CIB W92 Procurement Systems Symposium*, 4-7 December 1994, Hong Kong. The Department of Surveying, Hong Kong University, Vol. 1, 27-33.
Abstract: Novation contract is a relatively new procurement method employed in the construction industry of Australia. However, it is not a new concept on an international scale. It is a contractual process which dates from Roman civil law. Novation is the principle where a contract in existence between two or more parties with a new contract being substituted for it either between the same parties or between different parties, the consideration mutually being the discharge of the original contract. In this system the client engages design consultants to create a suitable design proposal. Before the design is fully documented, the proposal is tendered out to selected contractors who work out the remaining design with the consultants and tender for construction. The winning contractor accepts responsibility for the full design and build of the project and is required to hire the client's design consultants. This paper examines the concept of novating a contract and its application in the construction industry. The novation procurement process and the risk allocation for the system will be analysed. A case study on a recently completed project using the novation contract system is also included in the paper.
Notes: An introduction to novation as a procurement method with no mention of its relative costs.

Chang, C-Y and Ive, G (2001) A comparison of two ways of applying a transaction cost approach: the case of construction procurement routes. *In: Bartlett Research Papers*, No. 13, 41pp.
Abstract: Whilst the importance of transaction costs in construction has been accorded pervasive recognition, the methodologies used to apply this concept to the analysis of the construction process have been divergent. One way starts from searching for quantifiable items of transaction costs and explores the link between these costs and procurement routes. This paper claims that the attempt to provide explanatory foundation for construction procurement behaviour by quantitatively measuring each important element of transaction costs in the construction process is highly unlikely to succeed, since the majority of the costs with comparative importance are fairly difficult, if not impossible, to estimate. A possible way to avoid this pitfall is to follow Williamson's methodology of comparative institutional analysis. In this context, we take this to involve operationalizing the theory by predicting, for transactions with defined attributes, ordinal differences in transaction costs between institutions (procurement routes), and thus under a "weak" profit maximizing assumption, to derive and test refutable hypotheses concerning the probability or relative frequency of use of each route.
Notes: Chang and Ive divide possible methodological approaches into the "indirect measurement approach" (IMA) and "direct measurement approach" (DMA). The costs to be measured are divided into two groups (see Table 2 from the report). The first (TC1) are the costs that reduce the total economic gain from the transaction (because they are costs). These are: (1) Costs of information collection and search costs (or "selection" costs), (2)

Cost of bargaining: this is more likely to be the function of a specific department (the equivalent of "estimating and pricing"), (3) Cost of measuring performance of agents or quality of products (the equivalent of "monitoring, inspecting and supervising"). TC1s are all likely to arise in the process, no matter what procurement route is followed. The other sort of transaction cost, TC2, (see col. 3 of Table 2) may or may not arise. Its incidence and magnitude are unknown. The first type listed is "legal fees and opportunity cost of delay". Both of these consume resources, as in TC1. The complexity of the situation is to some extent shown in Figure 2 from the report.

—— **(2002)** On the economic characteristics of construction procurement systems. *In:* Lewis, J M (Ed), *CIB W92 Procurement systems*, January 2002, Trinidad. The Engineering Institute, University of the West Indies, St Augustine, Trinidad & Tobago, Vol. 1, 23.

Abstract: This paper examines the economic characteristics of procurement systems mainly in terms of transaction cost economics. This is an important founding block for a positive theory of construction procurement behaviour. This paper puts forth the theoretical principle of an inconsistent trinity, according to which, to determine the most efficient procurement method, the client will inevitably face the trade-off between (1) fastest delivery of the project, (2) high flexibility in accommodating changes (lower transaction costs arising from process specificity), as well as (3) single point of responsibility for design and construction (lower transaction costs from measurement problems).

Notes: A continuation of Chang and Ive's work in establishing procurement costs in a theoretical transaction cost context.

Chang, C-Y and Ive, G (2002) Rethinking the multi-attribute utility approach based procurement route selection technique. *Construction Management and Economics*, **20**(3), 275-84.

Abstract: The multi-attribute utility approach (MAUA) provides a possibility of transforming subjective perception or evaluation into objective decision principles. However, as applied to procurement system selection, the advice derived from this approach can be problematical if the nature of the procurement systems has not been examined carefully. The major weaknesses of the MAUA as applied in this field lie in its selection of priority variables and in some association of procurement routes with priority variables. This paper uses a transaction cost perspective, in a setting of incomplete contracting, to develop these comments.

Notes: Further useful theoretical work (transaction cost perspective) but with no aspiration to any empirical work at this stage.

Charoenngam, C and Sriprasert, E (2001) Assessment of cost control systems: a case study of Thai construction organizations. *Engineering, Construction and Architectural Management*, **8**(5/6), 368-80.

Abstract: The most important function that facilitates construction organizations to accomplish profit maximization is cost control. However, the absence of a well-established cost control system has caused failures to many Thai contractors, especially during the current economic recession period. To comprehensively understand cost control systems in practice, this study theoretically assessed effectiveness as well as the deficiencies of the traditional systems vis-à-vis the effective systems. In addition, by contrasting what are found in effective systems but not found in traditional systems, critical attributes most contributing to the systems' successfulness were identified so that the improvement steps can be suitably prioritized. The validated findings indicated remarkable contrasts between

traditional and effective systems. Two critical aspects, including advancement of cost control framework and systematic participation of site personnel in cost control, were found to be the major differences. Interestingly, similar conditions were encountered in various countries such as Greece, Pakistan and Australia; hence suggestions from this study could be internationally useful.
Notes: Concerned with monitoring, reporting and controlling actual production costs, rather that those of the commercial transaction.

Chau, K W and Walker, A (1994) Institutional costs and the nature of the sub-contracting in the construction industry. *In:* Rowlinson, S (Ed), *CIB W92 Procurement Systems Symposium*, 4-7 December 1994, Hong Kong. The Department of Surveying, Hong Kong University, Vol. 1, 371-8.
Abstract: This paper introduces an approach that is potentially very powerful in explaining observed phenomena related to the organization of construction activities. The approach is used to analyse one common phenomenon in the construction industry, i.e. sub-contracting. The pervasiveness of sub-contracting makes its efficiency fundamental to all procurement systems. Our preliminary findings suggest that the major reason for the widespread use of sub-contractors stems from the fact that the costs of identifying and agreeing prices of the components (the sub-contracting package) of a construction project are in general relatively cheaper than the cost of planning and monitoring workers' performance in a construction site. Under certain situations (such as when the nature of the work is uncertain or cannot be easily identified) where the former is more expensive, other forms of contract such as the wage contract will be adopted instead. Other explanations such as fluctuating workload, risk sharing and specialization, cannot be sustained. The argument that sub-contracting leads to exploitation and inefficiency is also found to be not sustainable.
Notes: An interesting theoretical statement that locates the decision to sub-contract in the context of transaction cost minimization.

Chege, L W and Rwelamila, P D (2001) Private financing of construction projects and procurement systems: An integrated approach. *In:* Duncan, J (Ed), *CIB World Building Congress*, 2-6 April 2001, Wellington, New Zealand. CIB9.
Abstract: Budgetary constraints in several countries have led governments to seek alternative methods for financing infrastructure provision. Public private partnerships (PPP) have received widespread attention in several countries in recent years. These PPP initiatives have enabled the public to utilize private sector finance and expertise for the provision of public infrastructure through schemes such as Design Build Finance and Operate (DBFO), Build Own Operate (BOO) and Operate and Transfer (BOT). This paper reviews these procurement systems and examines the relationship between these procurement systems and the financing of the project. One of the aims of utilizing these alternative procurement routes is to enable the client to obtain value for money and it is suggested that selection of an appropriate procurement system for a project will assist clients in attaining their objectives regarding the financing of the project. This paper reviews various PPP initiatives and utilizes the South African experience of the procurement facilities through PPP as a case study reflecting the financing of these types of projects. This forms part of a broader study geared at developing a methodology for the selection of appropriate procurement systems within Southern Africa.
Notes: A description of the PFI/PPP experience in South Africa.

Chen, J J (1997) The impact of Chinese economic reforms upon the construction industry. *Building Research and Information*, **25**(4), 239-45.
Abstract: The increase in economic activity in China has generated and will continue to generate a heavy demand upon the construction sector. The author studies the impact of the economic reforms upon the construction industry and the consequences of fulfilling an overheating demand. The new types of construction companies and ventures in China are identified and analysed. Both local and international private sector construction companies are needed to tackle the burgeoning demand.
Notes: The rapid increase in demand for construction in the Chinese construction industry has led to the development of competitive tendering. Previously, contractors in China were allocated work, as part of the machinery of a centrally planned economy. With the emergence of competitive tendering, the hope is for more efficient construction and less cost overruns. Moreover, it is felt that the only way that demand can be met is through international tendering, bringing in foreign contractors. This will also bring in advanced technology and modern management practices. While these conclusions are clearly stated in the paper, there is no research here to suggest that efficiency or technology transfer will result from the use of competitive tendering.

Chinyio, E A, Olomolaiye, P O, Kometa, S T and Harris, F C (1998) A needs-based methodology for classifying construction clients and selecting contractors. *Construction Management and Economics*, **16**(1), 91-8.
Abstract: Clients' needs are inadequately evaluated in project schemes. Without a precise establishment of clients' preferences, the essential criteria for project implementation and especially contractor evaluation cannot be appreciated fully. It is not surprising therefore that subjective decisions have prevailed in tender evaluations and clients' needs have not been satisfied completely. Objective contractor evaluation will be realized only when clients' needs and contractors' capabilities can be quantified and matched reliably. The methodology described herein rests on identifying clients' needs preferences as comprehensively as possible at project inception and progressing to identifying contractors who can satisfy them optimally; relying on "multidimensional scaling" and "cluster analysis" techniques. The investigations suggest that clients' project needs are not along the traditional classifications of private, public and developer clients; a reclassification of clients into five needs-based groups is proposed. A new contractor evaluation methodology matching client satisfaction to attainment of established needs in project schemes has been developed for bidding purposes.
Notes: Speculative study producing theoretical classifications of clients to provide a basis for objective decisions about contractor selection. No consideration is given to the relative importance of subjective decisions. Nothing about the cost of the process.

Christian, J (2002) A comparison of construction contractual arrangements. *In:* Lewis, J M (Ed), *CIB W92 Procurement Systems*, Trinidad. The Engineering Institute, University of the West Indies, St Augustine, Trinidad & Tobago, Vol. 1, 10.
Abstract: There are various types of contract in the construction industry, which include design-build, traditional, and partnering contracts. Partnering contracts are used so that a client, an engineer and a contractor can form a mutually beneficial contractual arrangement to complete a construction project. In the past few years, many privately funded partnering contracts for infrastructure projects have been completed. This type of contractual arrangement is often part of a government procurement policy to create a build-own-

operate-transfer arrangement. There are several public-private procurement strategies and systems available to the client. The paper reviews and examines the public-private procurement (PPP) systems available and comments on different contractual arrangements. The paper also describes research that investigated the effects of change orders on different types of contract. Construction project information and data were collected from a total of 40 public sector projects in New Brunswick. A local contract is described where the arrangement was changed from a privately financed contract with tolls to a contract without direct tolls.
Notes: Interesting examples of PPP in Eastern Canada. No details on the cost or relative cost of bidding on this method.

Chudleigh, J (1994) EC procurement legislation. *Architects Journal*, **200**(24), 46-7.
Abstract: The advent of the single market in Europe has increased opportunities in the public sector purchasing, but methods of bidding still vary. Looks at changes in tendering procedure following Directive 92/50/EEC.
Notes: A topical magazine article about the impact of EU procurement legislation in the early 1990s.

Clark, P (2001) Contactors on the defensive after PFI comes under fire. *Building*, **266**(26), June 29, 13.
Abstract: A report by the left-leaning Institute of Public Policy Research said the government should sometimes use different forms of procurement, and criticized some PFI schemes in the education and healthcare sectors. The publication of the report coincides with the announcement that nine hospital projects have been put on hold while a pilot scheme is assessed. This scheme involves ancillary staff remaining as NHS employees rather than being transferred to PFI consortiums. The IPPR report claimed that savings in hospital and school PFI schemes were minimal and called for more public–private partnership models, particularly those that did not include private financing.
Notes: Contractors and consultants defend the private finance initiative after an attack from a leading UK think tank.

—— **(2002)** PFI bidders deterred by costs, says report. *Building,* **267**(46), 22 November.
Abstract: The number of bidders for PFI schemes will drop because of the falls in the share price of contractors and the introduction of tougher accounting standards for bid costs, according to a report published. Just under 90% of respondents to the International Project Finance Association's PFI survey said there was less interest in PFI work. Industry fears centre on the fact that PFI firms must account for bid costs earlier than before and will be less able to obtain credit because of declines in their share prices. IPFA director Richard Kenton said the survey results showed that the government ought to act. The survey, which canvassed views from 223 PFI funders, contractors and advisers, also found that respondents were worried about union opposition to PFI – 65% said that unions would continue to be an issue for PFI next year. More than four out of five thought the role of government PFI adviser Partnerships UK was ill defined. In addition, the survey found that 97% had been asked to pay larger insurance premiums.
Notes: The IPFA felt that more needed to be done to lower PFI bidding costs, "which continue to be a burning issue for the market, especially when design costs are added."

Clarke, L and Wall, C (2000) Craft versus industry: the division of labour in European housing construction. *Construction Management and Economics*, **18**(6), 689-98.
Abstract: Two distinct divisions and concepts of labour are apparent from an analysis of social housebuilding sites in the UK, Germany and the Netherlands: the craft form, based on controlling the output of labour; and the industry form, based on the quality of labour input. These are associated with different work processes, skills and training, and also different levels of mechanization and component prefabrication. In the UK, which is craft-based, low levels of mechanization and prefabrication were found compared with Germany and the Netherlands, and the range of activities for the separate trades in assembling superstructure elements was simpler. Labourers are distinct from craftsmen and remain a significant group. Skills are narrow and training provision low. A high proportion of the labour force remains self-employed, under labour-only sub-contractors, working to price or output. In comparison, in Germany and the Netherlands labour is employed directly and work processes are more complex, with more specialisms at the interfaces. The division of labour is industry-wide, training provision is extensive, and skills are broad and integrated into the grading structure. Greater speed, higher productivity and lower levels of supervision are associated with industry-wide systems compared with traditional craft forms.
Notes: The incidence, in the UK, of labour-only sub-contracting is highlighted as an important aspect of the supply chains that are present in most construction projects.

Coase, R H (1937) The nature of the firm. *Economica*, **4**(16), 386-405.
Abstract: None
Notes: This is one of the seminal, early works by Coase that introduced the theory of transactions costs. Coase's original suggestion, developed by Williamson (q.v.) is that a transaction cost approach is the most appropriate way of accounting for the co-existence of markets and hierarchies as different organizational forms.

—— (1960) The problem of social cost. *Journal of Law and Economics*, **3**, 1-44.
Abstract: This paper is concerned with those actions of business firms which have harmful effects on others ... The economic analysis of such a situation has usually proceeded in terms of a divergence between the private and social product of the factory, in which economists have largely followed the treatment of Pigou in The Economies of Welfare. ... It is my contention that the suggested courses of action are inappropriate, in that they lead to results which are not necessarily, or even usually, desirable.
Notes: Coase's first work in the area of transaction costs was in 1937 but almost 30 years passed before he revisited the subject in this work. According to Coase (1960: 22) transaction costs arise from the need "to conduct negotiations leading up to a bargain, to draw up the contract, to undertake the inspection needed to make sure that the terms are being observed, and so on."

Commission on Public Private Partnerships (2001) *Building better partnerships: the final report of the Commission on Public Private Partnerships*, London: Institute for Public Policy Research.
Abstract: The starting point for the work of the Commission was the need to renew support for universal, public funded services through improvements in their quality and responsiveness. The aim was to assess the appropriate contribution of PPPs in delivering this agenda. Under the 1997-2001 Labour government PPPs have been the source of huge controversy: critics argue that they are little more than privatization by stealth, while

supporters maintain that PPPs are the key to the transformation of the UK's public services. This report cuts through the different economic and policy arguments around PPPs highlighting both the promise and the pitfalls of different models of partnership.
Notes: Among the points of interest is the statement that "all PFI projects are publicly funded and incur future liabilities for the exchequer". PFI seems to be offering significant gains (in value for money) in roads and prisons but not in hospitals and schools. Government should experiment with a range of procurement models for capital projects. A new mono-culture of procurement based on the current PFI model should be avoided. This is of background interest only as it deals with the long-term economics of PFI/PPP rather than their procurement costs.

Connaughton, J N (1994) Value by competition: a guide to the competitive procurement of consultancy services for construction. *In: CIRIA Special Publication*, No. 117, London, 51.
Abstract: This guide sets out the principles of good practice, and covers selection on the basis of consultants' abilities, their fees or both. While it is for the user to apply these principles in particular circumstances, the guide will help to: identify when competition is appropriate; understand the implications of competition; manage the competitive process, in order to select the most suitable consultant for the job. The guide concentrates on the better known construction consultancy services – architecture, engineering, surveying and project management. However, the principles discussed apply to the procurement of all consultancy services for both building and civil engineering work. There are two main ways of selecting and appointing consultants. Either a single consultant may be approached to discuss and agree the appointment, or a number of consultants may be invited to compete for it. This guide is concerned with the use of competition to select and appoint professional construction consultants. Competition is when two or more consultants formally and explicitly compete to provide their services. It includes, but is by no means restricted to fee competition. Design competitions are a special case and are not covered in this guide.
Notes: Of some background interest to the present work, but with no empirical evidence of the relative cost of the main ways of selecting and appointing consultants.

Construction Industry Council (1993) *The procurement of professional services: guidelines for the application of competitive tendering*. London: Thomas Telford.
Abstract: The Construction Industry Council states that the balance between quality and price should remain the sole basis of the final selection. Two systems of tendering are highlighted, both of which can be operated within the recommendations set down in the Treasury's Central Unit on Purchasing Guidance Notes, the DoE guidelines for the Design of Government Buildings and the Public Services Contracts Regulations (SI No 3228 which implemented in January 1994 the EC Services Directive). The first is a "Two Envelope" system where competing tenderers submit their offers in two separate envelopes. The first envelope is clearly marked "Technical Proposals", and should contain all the required technical content of the proposal such as method statement, programme and staff curricula vitae. The second envelope contains the fee. Only the first envelopes are opened at the outset of the tender appraisal. All the technical proposals are compared and placed in order of preference. Then, and only then, the fee tender of the first choice is opened. If it is found to be satisfactory, the commission is offered to that tenderer. The CIC recommend the "Value Assessment" system. In this system the selected quality criteria, as well as the price, are given pre-determined weightings to be applied by the tender evaluation panel. The information sent to the tenderers should include details of the weightings so that tenderers

can gauge the relative importance of the requisite quality criteria and submit their tenders accordingly.
Notes: Of limited interest to the present work in its description of two systems of tendering that maintain a balance between quality and price.

Construction Industry Institute (1991) *In search of partnering excellence*. CII special publication 17-1, Austin, USA: Bureau of Engineering Research, The University of Texas.
Abstract: In 1987, an industry association in the USA, the Construction Industry Institute (CII), established a task force to search out and structure a culture based upon Partnering. In Search of Partnering Excellence summarizes the Task Force's findings. The work sets down the basic precepts and methods of partnering.
Notes: Has some background relevance to the process of and results from partnering.

Construction Industry Research and Information Association (1998) *Selecting contractors by value*. London: CIRIA.
Abstract: The guide is addressed to all those within client organizations who are responsible for construction projects and to the consultants and others who advise them. It covers the whole process of selection by value, from the inception of a project to the point of starting construction on site. Rather than prescribe universal procedures, it is designed to help readers develop criteria and procedures that are appropriate for particular circumstances. The principles expressed in this guide are compatible with the EC procurement rules.
Notes: A procedural guide to contractor selection.

Construction Round Table and Business Round Table (1994) *Review of procurement methods*. London: Business Round Table.
Abstract: This up-to-date review of the main conditions of engagement and standard forms of contract for construction assesses their effectiveness in use in promoting of cost reduction, quality improvement, faster speed, and greater reliability of completion. It sets out a basic performance specification for a family of compatible and mutually reinforcing contracts for consultants, managers, works contractors and suppliers of modules and components. Prepared by the Construction Round Table of leading customers, currently buying over £5 billion a year from the construction industry, and based on their own experiences, the procurement methods described are intended to achieve improvements in performance while benefiting all parties.
Notes: A practical guide for clients to selecting the appropriate procurement method.

Cook, A (1999) If Laing lost, who wins? *Building*, **264** (44), November 5, 22-3.
Abstract: Analysis of the construction firm, Laing's, decision to move out of competitive tendering. Laing is now focusing on private finance initiative (PFI) and other negotiated work.
Notes: It is interesting that the larger contractors are moving towards negotiated work with a longer, more complicated and (possibly) more expensive selection procedure.

Cook, A E (1990) The cost of preparing tenders for fixed price contracts. *In: Technical Information Service*, No. 120. Ascot: Chartered Institute of Building.

Abstract: The procedures adopted for obtaining and analysing the tendering costs incurred by a number of contractors are discussed. This is followed by an explanation of the way in which the tender costs sampled are used to ascertain the overall cost to the industry. The significance of these costs is used as a basis of assessing the appropriateness of the tendering process. A comparison of tendering costs for different sized contractors is also undertaken. Alternative tendering procedures are regarded as viable and it is concluded that there is scope for a reduction on tendering costs.
Notes: An early and modest attempt to measure the actual costs of tendering. It is argued that it is an inherently costly and time consuming process for contractors and their clients especially with abortive tendering. This proposition is tested by determining and analysing the significance of such costs, using sampling techniques to ascertain the total annual tendering costs for the industry. The appropriateness of the tendering process which gives rise to such costs is questioned.

Cowan, R (1987) The tender trap. *Architect's Journal*, **186**(44), 26-9.
Abstract: "Competitive fee tendering is a lottery where everyone can lose". Robert Cowan investigates what can be a perfect way to waste a lot of time.
Notes: Competitive fee tendering may produce a large number of tenderers. The author complains about the waste of time and money for both the Council and tenderers. One tenderer estimated that the cost of preparing the tender was £2000 (time of senior staff being valued at £30/hour). Assuming all the tenderers put in the same amount of resource, the total sum would be £50,000.

Cox, A W and Ireland, P (2002) Managing construction supply chains: the common sense approach. *Engineering, Construction and Architectural Management*, **9**(5/6), 409-18.
Abstract: The paper provides an introduction to the argument that there is considerable evidence of poor thinking within the construction industry. The failure to understand the circumstances that are facing industry players will prevent clients, contractors and suppliers from achieving their own objectives. The current problems are further compounded by the advice espoused by the government-sponsored industry reports advocating generic approaches. In response to these problems the paper provides practitioners with a theoretical framework for understanding: the structure of the industry and its constituent supply chains; the attributes of buyer and supplier power; the appropriateness of certain relationships according to the firm's power position within the construction supply chain; and, the circumstances where the recent industry initiatives and an integrated supply chain approach may be implemented with success.
Notes: A descriptive paper, focusing particularly on supply chains and relevant only in terms of context.

Cox, A W and Thompson, I (1998) *Contracting for business success*. London: Thomas Telford Publishing.
Abstract: Taking the balance of power and allocation of contractual risk as its theme, this book provides a comprehensive overview of all the major forms of contract and procedures for dispute resolution used in the construction process. The result of long-term industry-funded research, it considers which forms of contract and which dispute resolution techniques are most appropriate in different circumstances and includes a model for selecting the optimal procedures. The use of partnering and supply chain management in the

construction industry is examined and the book concludes with a guide for selecting the best methods.
Notes: The transaction costs of construction contracting are mentioned in several places. They are identified as "the identification, accreditation and selection of suppliers, and/or performance monitoring and feedback" which would not necessarily apply if the activity were carried out in-house. It is noted that there are considerable transaction costs associated with bespoke contracts or with newly-drafted contracts. The authors argue that "managed collaboration with a limited number of suppliers in competition saves transaction costs".

Cox, A W and Townsend, M (1998) *Strategic procurement in construction: towards better practice in the management of construction supply chains*. London: Thomas Telford.
Abstract: In addressing the largely unrecognized problem of misalignment between business strategy, supply relationships and operational practices in construction, this book explores how appropriate certain procurement strategies are to given situations.
Notes: The book contains (Section B) six procurement case studies. Chapter 13 (Strategic Cost Management) alludes to the elimination of unnecessary cost: some benefits can be had from different approaches to procurement. For example (citing a report by the Construction industry Institute in the USA, *q.v. infra)* "where a high degree of trust exists between parties ... cost benefits are likely to result".

Craig, R (1999) How innovative is the common law of tendering? *Journal of Construction Procurement*, **5**(1), 15-26.
Abstract: Traditional design-by-owner remains an important procurement option despite the advances made by design-build in recent years. Contractor-led innovation is important and desirable in both procurement options; yet traditional design-by-owner procurement processes prevent, restrict or even discourage such innovation. Developments in common law are revealed which result in contractual obligations for the procurer which might further inhibit innovation, as the procurer becomes obliged to treat all tenderers equally and fairly. The theory of the tendering contract is introduced and the problems for the procurer discussed when presented with a non-conforming alternative tender that offers a significant cost-saving against conforming tenders. However, if accepted, this puts the procurer in breach of contract to at least one aggrieved tenderer. The conclusion is reached that in order to properly consider alternative tenders without failing in its obligation to treat all conforming tenderers equally and fairly, the owner must make specific provisions within tender conditions which create the power to consider alternative proposals.
Notes: The main relevance of this paper to the present study is its argument that tendering malpractice could be eliminated or reduced by the concept of a "tendering contract" that has a limited constraint on both procurer and prospective supplier.

—— **(2000)** Re-engineering the tender code for construction works. *Construction Management and Economics*, **18**(1), 91-100.
Abstract: The NJCC Code of Procedure for Single Stage Selective Tendering (now withdrawn) and its successor, the Construction Industry Board's Code of Practice for the Selection of Main Contractors are criticized in the light of decisions of the common law courts with regard to regulation of the tendering process. It is argued that a new "Tendering Code" should be produced in the style and format of a contract document which reflects not only the statutory regulation imposed on public bodies, but common law decisions of the courts. The nature of this tendering contract is explained as a means of regulating the

tendering process. Issues discussed are: dealing with errors and irregularities found in tenders; dealing with non-compliant tenders; dealing with tender withdrawal prior to its acceptance or rejection; making provisions as to time for submission of tenders and dealing with late tenders; making provision for submission of tender by fax or other electronic means; making provision for evaluation of tenders received; and imposing or negotiating reductions in price with tenderers prior to acceptance. The paper concludes that the common law obligations placed on the owner to treat all tenderers equally and fairly and to apply the tender conditions when evaluating tenders and awarding contracts seems to be good common sense and of commercial advantage, not only to the immediate parties concerned but also to the wider community.
Notes: This paper discusses the provisions of the old NJCC tender code documents and also the successor CIB documents. It recommends their replacement by the standard form of contract which takes into account some of the problems discussed.

Crowley, L G and Karim, A (1995) Conceptual model for partnering. *Journal of Management in Engineering, ASCE*, **11**(5), 33-9.
Abstract: Partnering is typically defined in one of two ways: by its intended attributes such as trust, mutual goals, long-term commitment; or by the process where partnering is seen as a verb, as in developing a mission statement, agreeing on goals, etc. These definitions of partnering illustrate the intended results of partnering and the process that brought them about; however, they leave undefined the entity of partnering. This definitional bias contributes to existing limitations in the implementation of partnering, such as the unpredictability of success versus failure in a partnering situation. In this paper, partnering is conceptually defined as an organization formed through the implementation of a cooperative strategy by modifying and supplementing the traditional boundaries that separate companies in a competitive climate. In this way, partnering wraps the major project participants into an alliance, creating a cohesive atmosphere for open interaction and cooperative project performance.
Notes: Has some background relevance to understanding partnering.

Dawood, N N (1995) An integrated bidding management expert system for the make-to-order pre-cast industry. *Construction Management and Economics*, **13**(2), 115-25.
Abstract: Bidding decisions, including the estimation of optimal mark-up on price, represent major decision problems for companies formulating a successful business strategy. The objective of this research was to develop an integrated bidding management expert system to assess the suitability of incoming enquiries for a particular company and suggest a "bid/no bid" decision. If a decision to bid is taken then the expert system should provide advice on the optimal mark-up to maximize the chance of winning potential contracts. The system developed in this paper is composed of an information system that integrates design, estimation and production planning, and a knowledge base to provide abstracted information and advice to managers in charge of bidding. The information system analyses the records of previous contracts, and present managers with vital information that minimizes the risk of poor decisions associated with bidding. The knowledge base is composed of intelligent rules which were elicited from previous contract records and experienced managers in charge of bidding. They are designed to advise managers on two major issues: bid/no bid and estimation of optimal tender price. A number of factors that affect bidding strategies were identified by reviewing previous bidding methodologies and surveying eight major companies in the UK by means of semi-structured interviews.

Notes: The concern of this paper is the optimization of mark-up to tenders, and it has therefore little other than background relevance to the present study.

Donovan, P (1997) Juggling price and quality. *Contract Journal*, **389**(6129), June 11, 12-13.
Abstract: The Government's plans to phase out Compulsory Competitive Tendering (CCT) in favour of the new Best Value practice has caused concern among many local authorities. A review of the proposed changes is being undertaken by the Government in consultation with the Local Government Association.
Notes: One of a number of papers discussing the move from CCT to Best Value procurement in the UK public sector. The point of interest is that this move was the single most important factor in enabling public sector procurers to contemplate the use of collaborative procurement arrangements.

Dorée, A G (1996) Tendering for co-operation municipal-contractor. *Heron*, **41**(4).
Abstract: The European Union (EU) has strongly influenced the perspective on the way public organizations (should) utilize markets. The EU directives on procurement, and the battle against cartels, resulted in extra attention towards the relationships between municipalities and construction contractors. Figures suggested that municipalities prefer limited tendering procedures, and avoid public tender procedures. This paper reports on the research into the reasons for such behaviour. Analysis of municipalities' procurement and tendering practice uncovered an intricate mechanism for project control. Municipalities implicitly use the prospect of future assignments to restrain contractors' misbehaviour. By doing so municipalities reduce uncertainties and risks. Contractors' demeanour becomes more flexible, co-operative and quality orientated because of this mechanism. Through the use of this mechanism the municipality-contractor relationship has developed to a kind of co-makership relation. Bending the procurement and tendering practice towards more public tendering is expected to make project control more troublesome.
Notes: The paper investigates the situation in the Netherlands where municipalities preferred limited tendering procedures to avoid public tender procedures. Thirty-five interviews were conducted and in addition 117 projects in 117 municipalities were investigated from the contractors' and clients' side. In essence, municipalities use a continuing relationship with contractors and the prospect of future contracts to restrain contractors' behaviour. The author relates this to transaction cost theory. The argument runs as follows: contracts are drafted to reduce risks but risks remain. If renegotiation is necessary post-contract, the contractor is in a strong position, being the only one involved. Thus, contractors may indulge in "opportunistic" behaviour and use their strong position to get more from the contract. They could also string the work out so it will not get noticed till after it is paid for. However, contracts cannot normally be stockpiled, and in *lieu* of a stock of contracts they try to get a stock of clients. They do this by being amenable and doing good work so that the continuing client will favour them in the future. At the same time the municipalities realize that they are at risk of "opportunistic" behaviour and like to employ contractors they know have been good in the past. In Williamson's terms they "introduce trading regularities that support and signal continuity intentions".

—— **(1997)** Construction procurement by Dutch municipalities. *Journal of Construction Procurement*, **3**(3), 78-88.
Abstract: Over the last decade the European Union (EU) has had a strong impact on the way public organizations introduce market philosophies in their policies. The EU directives

on procurement, the battles against cartels, the globalization and harmonization of markets, all lead to extra attention to the procurement practice of public agencies. The procurement practice of one of these public organizations, the municipalities in the Netherlands, is examined. Statistical data suggest that Dutch municipalities prefer limited tendering procedures, and seem to avoid public tender procedures. Research into the reasons for such preference is reported. Analysis of municipalities' procurement and tendering practice uncovered an intricate mechanism for maintaining project control. Municipalities implicitly use the prospect of future assignments to restrain contractors' misbehaviour. By doing so, municipalities reduce uncertainties and risks. Through the use of this mechanism the municipality-contractor relationship has developed to a kind of co-makership relation. This phenomenon is categorically overlooked in the standard market paradigms.
Notes: A similar message to the previous paper, extended to consider mechanisms for project control.

—— **(2004)** Collusion in the Dutch construction industry: an industrial organization perspective. *Building Research and Information*, **32**(2), 146-56.
Abstract: Several investigations by parliament, cabinet, justice and antitrust authorities have shown a widespread use of cartels and structural bid rigging within the Dutch construction industry. The reputation of the Dutch construction industry has been dented with both the general public and clients. As a response, the Netherlands' Parliamentary Inquiry Committee on Construction Fraud adopted the guiding principle of "competition is good" and urged the restoration of the proper functioning of the market. The proposed default approach to public sector procurement is design-bid-build with public tendering and selection of the lowest price. A concise overview of the investigations is provided, relating the collusions and their persistence to emerging insights from the field of industrial organization theory into underlying factors and causes. A tougher public sector procurement policy and the continued reliance on lowest bid prices may not contribute to the reform of the Dutch construction industry as intended. One-dimensional, price-oriented competition only provides a static, project-based efficiency. However, it neither addresses a number of organizational issues nor resolves the underlying pressures leading to collusion. An alternative approach allowing for a balance of competition and collaboration with a wider number of selection criteria variables would create a more dynamic, iterative competitive process over a longer timeframe and would develop an innovative, efficient and profitable industry. Although the inquiry committee acknowledges these new methods of procurement, it is expected that the overriding ambition to restore proper market function (through increased competition) will steer towards the more traditional procurement approaches.
Notes: A paper that presents a reminder that the abandonment of price competition can go too far, and transgress the law. The conclusion addresses the need for a balance between competition and collaboration.

Dorée, A, Holmen, E and Caerteling, J (2003) Co-operation and competition in the construction industry of the Netherlands. *In:* Greenwood, D J (Ed), *19th Annual ARCOM Conference*, 3-5 September 2003, University of Brighton. Association of Researchers in Construction Management, Vol. 2, 817-26.
Abstract: In 2002, a Parliamentary Enquiry Committee exposed widespread collusion practices in the Dutch construction industry. The construction industry in the Netherlands is in turmoil and is seen as not living up to the standards that society requires. There seems to be a culture and an environment that induces and sustains economic offences and

malpractices. Furthermore, performance and progress are not keeping pace with other industries. The Parliamentary Enquiry Committee recommended a rethinking construction type of reform initiative. In several countries around the globe, there are ongoing efforts to reform construction. International evidence, as presented at the "revaluing construction" conference in February 2003 in Manchester (CIB/UMIST), shows a trend towards more co-operative relationships and integrated procurement. Pivotal in all reform initiatives are changes in the approach towards public sector procurement, with public clients acting as leading clients in the reform. The aim of the project presented in this paper is to study construction industry reform from the theoretical perspective of industrial organization economics and market dynamics theory. The project connects the construction industry (reform initiatives) with the emerging theoretical insights on dynamic competition. The construction industry might be a good example to test the newly emerging insights in the field of industrial economics. A better understanding of the business systems and market structures, and of lessons learned abroad, may substantially improve the chances of successful reform of this troublesome industry.
Notes: A continued examination of the issues considered in the previous paper.

Dorrell, E (2002) MoD drops bombshell over its difficulties with PFI schemes. *Architects Journal*, **216**(19), 21 November, 10.
Abstract: The Ministry of Defence has admitted that nearly 50% of its PFI schemes have either failed to come in on budget or have run over time. Despite attempting to put a positive spin on a new report's findings by highlighting the successful schemes, the government department has conceded that its conclusions are "disappointing". The survey which included several large-scale construction projects, such as the Army Foundation College in Harrogate and the Joint Services Combined Staff College in Cranfield, found serious delays in six out of the 19 schemes – or 31% of them. The survey will also come as a hammer blow to those PFI supporters who argue the procurement method forces contractors to hit their financial targets. Only 10 out of the 19 projects came in on budget. And of the remaining nine projects, six had undisclosed additional costs that were "more than insignificant". Armed Forces minister Adam Ingram admitted that PFI schemes have had their teething problems. "There will of course be some projects where we need to work with the partners to improve", he said. "It would be foolish to think that everything always goes right". But Ingram insisted there are positives to take from the survey and said the government is committed to see the scheme work. "The overriding message is that Private Public Partnerships can, and are, delivering better services that we seek", he added.
Notes: Of little relevance to the present study, save as an example of one of the many discussions of the pros and cons of PFI.

Dozzi, P, Hartman, F, Tidsbury, N and Ashrafi, R (1996) More-stable owner-contractor relationships. *Journal of Construction Engineering and Management, ASCE*, **122**(1), 30-5.
Abstract: This paper presents an updated and enhanced version of a study conducted by the task force of the Construction Owners Association of Alberta (COAA) on "More Stable Contractor Relationships". The purpose of the study was to explore the current contracting philosophies, strategies, methods of execution, tendering process, and techniques, to identify the problem areas and determinants of success in the execution of projects, and to share these findings with the industry for further improvement. We have found that the industry is somewhat complacent and appears reluctant to embrace change. There is a preoccupation with contract award on the basis of lowest lump-sum bid although most recognize that this

often results in a higher end price for the product. There is a need for improvement in the areas of life-cycle costing, contracting process, risk management, teamwork, trust, and cooperation, as well as communication and the use of quality management principles. There are ample opportunities to be more progressive. The survey and subsequent evaluation concluded that any initiatives to formulate more stable contractor relationships must come from construction owners.

Notes: Through a survey of 16 construction-related firms, owners, contractors and consultants, the stability of owner-contractor relationships was explored. A focus on lowest price fails to exploit constructability, alternative methods and materials, and teamwork. They report that project success is more closely related to functionality and schedule than it is to cost. The paper is confusingly presented in that citations to the literature are placed with the discussion of responses, where the literature supports what the researchers are discovering. There are frequent references to an unpublished survey and questionnaire undertaken by the authors. Major findings are that preoccupation with lowest price bidding often results in a higher end-price for the product and prevents the contractor from making valuable contributions to the design. Interesting quote: "Most of the respondents feel that business environment and the project specifics affect the selection of the contracting approach significantly." (p32). For new technology, demolition and renovation, owners tend to assume more risk. The business environment affects the aggressiveness of clients in their transfer of risk. Specifically, in a buyers' market, owners will push risk on to contractors more aggressively than in a busy construction market. Basically, the paper makes suggestions in line with much of the thinking connected with improving contract practices in the industry.

Duff, R, Emsley, M, Gregory, M, Lowe, D and Masterman, J (1998) Development of a model of total building procurement costs for construction clients. *In:* Hughes, W P (Ed), *14th Annual ARCOM Conference*, September 9-11, University of Reading, UK. Association of Researchers in Construction and Management, Vol. 1, 210-18.

Abstract: There is a dearth of information on the comparative costs of projects carried out using the main procurement building systems. This paper reports the feasibility study of a research programme to produce a computer-based neural network cost model to show the effect on client costs of, *inter alia*, using different procurement approaches. A literature search identified 39 cost-significant project variables. Data were collected from collaborating QS practices, resulting in 46 project data-sets with which to test various modelling approaches. Evaluation of the data and model objectives identified multiple regression and neural networks as potential model forms. Investigation and trials of both have shown that regression and neural networks can provide effective representation of the client costs model but neural networks, due to their greater ability in modelling interdependencies between input variables, modelling non-linear relationships, and handling incomplete data sets, will probably be the better choice with which to analyse the very much larger volume of data planned for the next phase of the research. The results have demonstrated that such a model can be developed, that data to support it can be obtained and, additionally, that the utility of the model may be significantly greater than had been envisaged at the start of the study.

Notes: This is of some relevance to the current project, though its concern is to produce a prototype model that predicts the effect of procurement method on *total* project cost, without distinguishing the transactional costs of the procurement process itself.

Dulaimi, M F and Dalziel, R C (1994) The effect of the procurement method on the level of management synergy in construction projects. *In:* Rowlinson, S (Ed), *CIB W92 Procurement Systems Symposium*, 4-7 December 1994, Hong Kong. The Department of Surveying, Hong Kong University, Vol. 1, 53-60.
Abstract: This paper reports the findings of a study into the construction project teams' level of integration and its relationship with procurement systems. One term used to explain this process is synergy. Two procurement methods were compared, these were the design and build and the traditional procurement approaches. The hypothesis is that design and build method of procurement can improve the management synergy of a construction project. The result seems to confirm the above hypothesis. It shows that communication has improved significantly between project team members.
Notes: Some background interest in its comparison of traditional and design-and-build approaches.

Dulaimi, M F and Shan, H G (2002) The impact of contractors' size on their attitude towards the factors influencing bid mark-up decisions in Singapore. *Construction Management and Economics*, **20**(7), 601-10.
Abstract: The construction industry in Singapore is dominated by a competitive business environment which is being driven by lowest cost mentality. The pressure on contractors' profit margins has further increased with a prolonged recession in this sector which has seen construction demand and output retracting significantly. This paper examines the factors that contractors perceive to be important when they are considering the size of their bid mark-up. The research hypothesis is that contractors' size would have a significant influence on the factors that would influence the bid mark-up decision. Forty different factors were identified and a survey was conducted. The results have shown that the size of contractors has a significant impact on their attitude towards bid mark-up decision making. The analysis has shown that when deciding the size of a bid mark-up, large contractors tend to be more concerned with the nature of the construction work while medium size contractors focus on the state of their own company's finance.
Notes: The focus of this paper is on bid mark-up decisions, rather than the cost of bidding, or any other transactional aspect of procurement.

Eccles, R G (1979) *Organization and market structure in the construction industry: A study of sub-contracting.* Unpublished doctoral dissertation, Harvard University, Cambridge, Mass.
Abstract: not available
Notes: Academic underpinning for Eccles' later publication (see Eccles, 1981).

—— (1981) The quasi-firm in the construction industry. *Journal of Economic Behaviour and Organization*, **2**, 335-57.
Abstract: Construction projects are executed by general contractors who retain the services of speed trade sub-contractors. This form of organization is preferable to vertically integrating these trades because of the transaction cost implications of construction technology. The general contractor and special trade sub-contractors can form a stable organizational unit when conditions permit. This organizational form, called here the "quasi-firm", is analogous to the "inside contracting system" discussed by Williamson (1975). This paper uses Williamson's (1975, 1979) transaction cost approach to argue the

theoretical existence of the quasi-firm in the construction industry. Empirical evidence from a field study of homebuilders is presented in support of this argument.
Notes: Eccles finds elements of transaction cost theory useful. He describes the governance structure he discovered in Massachusetts among 26 small general contractors and their sub-contractors as intermediate between market and hierarchy; a "quasi-firm". Such work is welcome as it reveals that projects involving a nexus of sub-contracts are more than just a multiplicity of separate transactions. In fact, Eccles is one of a number of authors who have replaced Williamson's (q.v.) original dichotomous view of institutions with a "continuum" model of exchange that ranges from hierarchies to markets via intermediate forms such as networks. This continuum model was ultimately acknowledged and accepted by Williamson himself (1985: 83).

Egan, J (1998) *Rethinking construction: the report of the Construction Task Force to the Deputy Prime Minister, John Prescott, on the scope for improving the quality and efficiency of UK construction.* London: Department of the Environment, Transport and the Regions Construction Task Force.
Abstract: (extracts from report summary) The UK construction industry at its best is excellent. Its capability to deliver the most difficult and innovative projects matches that of any other construction industry in the world. Nonetheless, there is deep concern that the industry as a whole is under-achieving. It has low profitability and invests too little in capital, research and development and training. Too many of the industry's clients are dissatisfied with its overall performance. The Task Force's ambition for construction is informed by our experience of radical change and improvement in other industries, and by our experience of delivering improvements in quality and efficiency within our own construction programmes... If the industry is to achieve its full potential, substantial changes in its culture and structure are also required to support improvement. The industry must provide decent and safe working conditions and improve management and supervisory skills at all levels. The industry must design projects for ease of construction making maximum use of standard components and processes. The industry must replace competitive tendering with long term relationships based on clear measurement of performance and sustained improvements in quality and efficiency.
Notes: The members of the Task Force and other major clients will continue their drive for improved performance, and will focus their efforts on the demonstration projects. The authors ask the Government and the industry to join in rethinking construction. The Egan Report ("Rethinking Construction") was commissioned in October 1997 by John Prescott, Deputy Prime Minister. Published in July 1998, the central message of the report is that through the application of best practices, the industry and its clients can collectively act to reverse its negative trends. The report has been seen as important in achieving cultural change within the industry. It set the construction sector some clear objectives (initially intended to be on a year-on-year basis):
20% reduction in capital cost and construction duration
20% reduction in defects and accidents
10% increase in productivity
20% increase in predictability of project performance.
An integral element of both Rethinking Construction and Agenda for Change is the reduction of waste throughout the process from design to operation, leading to increased profitability.

Emmerson, J (1993) Competitive tendering and the Lincolnshire highways experience. *Proceedings of the Institution of Civil Engineers-Municipal Engineer*, **98**(1), 25-9.
Abstract: Lincolnshire County Council has adopted a strategy for competitive tendering which is well in advance of statutory requirements. Implementing the strategy has led to the introduction of competition into highway maintenance and construction, grounds maintenance, waste disposal, materials testing and geotechnics, and professional highway engineering services. The consequential organizational change has resulted in a much smaller Highways and Planning Department with the overall establishment reduced from 950 to 298. The strategy has produced substantial net savings in the cost of both works and professional services, but there have been problems due to inflexibility and delays in the procurement process. In addition serious problems have emerged in the quality of some professional engineering services, albeit in a minority of cases. Looking to the future, a more imaginative approach is needed to the procurement of local services including the development of long-term client/contractor partnerships and more emphasis on quality and completion times.
Notes: Summarizes Lincolnshire County Council's new tendering strategy for its roads.

Farrow, J J and Fraser, R (1996) Use and abuse of the Code of Procedure for single stage selective tendering. *In:* Harlow, P (Ed) *Construction Papers*, No. 70, Ascot: Chartered Institute of Building.
Abstract: This paper presents the findings of a survey of the views of professionals and contractors on the implementation and use of the Code of Procedure. This report provides an insight into the problem areas. The procedure advocated by the Code contains faults both in principle and application. It is recommended that the code should indicate to the contractor how many tenderers will be invited, thus the tenderer can make a better assessment of their chances and abortive tendering costs. It concludes that consideration should be given to the scope of the Code being extended to cover all aspects of construction activity.
Notes: Of relevance in the areas of excessive competition and abortive tendering costs. The NJCC Codes of Procedure for tendering have subsequently been replaced by those of the Construction Industry Board (Construction Industry Board, 1997b and 1997c).

Fayek, A, Ghoshal, I and AbouRizk, S (1999) A survey of the bidding practices of Canadian civil engineering construction contractors. *Canadian Journal of Civil Engineering*, **26**(1), 13-25.
Abstract: This paper presents the findings of a survey of the bidding practices of Canadian civil engineering construction contractors. The results of the survey provide insight into the most important factors that contractors consider in making four bid decisions: the decision to bid, the risk allowance, the opportunity allowance, and the mark-up size decision. The survey methodology is described to illustrate its effectiveness. Common practices in assessing risks and opportunities, the competition, and mark-up are discussed. A major conclusion of this paper is that the decision-making process used in bidding is largely subjective and based on experienced judgement. The assessment of the competition is done on an informal basis in most cases, with little use of historical competitor data. Risk and opportunity assessment is subjective and largely based on experience. Although the mark-up size decision is critical to the success of a company in achieving its objectives and realizing a profit, mark up setting is usually based on experience, with little or no formal methods of analysis.

Notes: The concern of this paper is the investigation of mark-up to tenders, i.e. price formation rather than cost assessment.

Fenn, P and Singh, H (1993) Recent trends in litigation in UK construction projects. *In:* Eastham, R and Skitmore, R M (Eds) *Procs 9th Annual ARCOM Conference.* Exeter College, Oxford, 14-16 September 1993, Vol. 1, 269-79.
Abstract: This paper describes the early stages of a research project funded by the Science and Engineering Research Council (SERC). The nature of the reported construction cases in the UK is examined and the level of activity in the specialist construction industry court, the Official Referees' Court, investigated. Suggested reasons for the apparent anomalies in empirical data are presented. A rise in summonses and interloculatory proceedings is investigated. The paper points toward further research which is ongoing.
Notes: The study hypothesizes about levels of litigation and levels of activity in the specialist construction sector.

Ferguson, N S, Langford, D A and Chan, W M (1995) Empirical study of tendering practice of Dutch municipalities for the procurement of civil-engineering contracts. *International Journal of Project Management*, **13**(3), 157-61.
Abstract: As European Union directives unify the method of tendering for European public-sector payouts, it is timely to review the tendering procedures formerly used by Dutch municipalities in the procurement of civil-engineering contracts. This historical review includes a consideration of the mechanism and effect of price improvement and tendering cost compensation on the overall tender price. A questionnaire survey was carried out, and several interviews with senior municipality officials were held in order to establish the factors which affected the choice of tendering procedure adopted, the criteria used for the award of a contract, and the desired quality of work. The principal conclusions reached were that some form of selection of contractors was preferred to straightforward public tendering, the reputation of a contractor was the single most important criteria when it came to selecting a contractor, and that negotiated contracts delivered a slightly higher quality of work than did either public or private tendering. This conclusion challenges the current directive which insists upon competitive tendering for all European Union funded work.
Notes: A critique of the rationale behind the European public-sector procurement directive, using empirical evidence from the tendering preferences of Dutch local authority clients.

Flanagan, R and Norman, G (1983) The accuracy and monitoring of quantity surveyors' price forecasting for building work. *Construction Management and Economics*, **1**(2), 157-80.
Abstract: The performance of two public sector quantity surveying departments is examined, when forecasting the lowest tender price for proposed projects at the design stage. The reliability of any price forecast is dependant upon professional skill and judgement and the availability of historical price data derived from completed projects. The quantity surveyor also requires an effective feedback mechanism that provides information on the accuracy of previous forecasts. A simple feedback mechanism is developed, which can be used to assess forecasting performance and give an early warning of bias and identify any patterns that may emerge.
Notes: More concerned with the accuracy of historical price data than the costs of procurement.

—— **(1985)** Sealed bid auctions: an application to the building industry. *Construction Management and Economics*, **3**(2), 145-61.
Abstract: This paper examines some of the theory and practice of competitive tendering in the building industry. Allocation of building contracts by means of competitive tendering is one of type of sealed bid auction. Two main strands of the theory of sealed bid auctions are developed. The first strand identifies optimal mark-up, given the tendering and resource constraints of tenderers. It emerges that the first strand is a special case of the second. A simple optimal bid price rule is identified, which indicates that bid price for any one contractor will be affected by the resources available to the contractor and general market conditions. It is shown that bid prices will be more competitive the better the information available to tenderers, the more are fully is the tender list constructed and the greater the number of firms invited to tender. It is further shown, however, that little is gained by having more than five firms on the tender list. Some empirical evidence is examined in the light of this theory and is shown to be consistent with it.
Notes: General background work on the theory and practice of competitive tendering in the building industry.

Gordon, C M (1994) Choosing appropriate construction contracting methods. *Journal of Construction Engineering and Management, ASCE*, **120**(1), 196-210.
Abstract: This paper examines the compatibility of various construction contracting methods with certain types of owners and projects. Contracting methods, as defined in this paper, consist of four parts: scope, organization, contract, and award. An owner must create an appropriate contracting method for each project. It was determined that there were six main organizations around which the contracting variations are created: general contractor, construction manager, multiple primes, design-build, turnkey, and build-operate-transfer. The most common method is the traditional system of an independent designer, a general contractor, and a competitively bid, lump-sum price. This method is efficient in many cases; however, in some situations, alternative methods are more appropriate and should be explored. Choosing certain methods can decrease the project duration, provide flexibility for changes, reduce adversarial relationships, allow for contractor participation in design, provide cost savings incentives to the contractor, and provide alternative financing methods. Guidelines are established to help the owner choose the organization, contract type, and award method most applicable for their project and themselves.
Notes: Discussion and categorization of available procurement options in the industry.

Graves, A, Sheath, D, Rowe, D and Sykes, M (1998) *The government client improvement study*. Bath: Agile Construction Initiative, University of Bath.
Abstract: It is clear that there is the potential for a significant step change in Government construction procurement practice. This report has focused upon those international and public/private sector comparisons that are relevant to central government construction and provide hard evidence for each potential improvement. Thus any significant change in procurement practice could be the most important factor in achieving positive change in the structure of UK construction. The route towards this improvement target is not unknown. Much good work has been done by Government, and more is currently ongoing, through demonstration projects such as the Construction Supply Networks Project funded by the MoD and DETR. However, there is a need to accelerate the rate of improvement. This requires sustained top-level support for an initial three-year programme focusing on clear targets for strategic change in Government construction procurement practice. A cross-

government body, such as the Government Construction Client Panel (GCCP), is well positioned to be the prime working mechanism for such a change programme. Partnering, framework agreements and other innovative approaches to procurement are all possible under current procurement rules and regulations, yet they are largely unused within the public sector. Management flexibility within the bounds of current regulatory frameworks should be promoted. Measurement systems should be developed to act as the "wake-up call" for those who do not yet see the need for improvement. Key performance indicators, at every level of the project process, are required. This requires a creative approach to project procurement. Procurement rules and regulations bound but do not restrict practice. Through more innovative approaches to procurement significant improvement is possible. The construction of the best Government client is achievable and should be developed.
Notes: This report pre-dates the move, in 2000, from CCT to Best Value procurement criteria in the public sector. Calls for more innovation in procurement methods.

Gray, C and Flanagan, R (1989) *The changing role of specialist and trade contractors.* Ascot: Chartered Institute of Building.
Abstract: This research study provides an insight into the structure, processes and relationships that relate to the diverse sub-contracting sector of the construction industry. Many pressures coupled with a reducing, but increasingly volatile workload, have forced a flexible form of working to emerge, encouraging the development of sub-contracting as has the increasing specialization of building technology and components which in turn has created a great diversity of manufacturing and supply organizations. To produce a building effectively requires the close co-operation of designers, specialist contractors and contractors all working to the same objectives, the management framework recognizing the contribution that each has to make in the complex, interconnected design and construction process through the contractual responsibilities of the parties involved.
Notes: A seminal report of the late 1980s that highlighted the importance, complexity and contractual difficulties of the specialist sub-contract sector.

Gray, C and Hughes, W P (2000) *Building design management.* Oxford: Butterworth Heinemann.
Abstract: A practical handbook on the management of building design, this guide explains the processes, roles and responsibilities of those involved in the design of the building, as well as ways to maximize efficiency. The book includes notes and checklists on, for example, how to select a design team and how to organize and plan the design process.
Notes: Some background relevance in terms of design team selection.

Great Britain Treasury Procurement Group (1997) *Procurement guidance.* London: HM Treasury Procurement Group.
Abstract: Responsibility for the formulation of integrated procurement policies and strategies throughout Government Departments rests with the Office of Government Commerce (OGC). OGC was set up on 1 April 2000 as a successor to the Treasury Procurement Group, the PFI Task Force, the Buying Agency (TBA), Property Advisers to the Civil Estate (PACE), and the Central Computer and Telecommunications Agency (CCTA). OGC is also responsible, among other things, for measuring and benchmarking procurement across Government, providing a centre of excellence for strategic procurement skills, such as PFI, outsourcing and the management of very large complex projects, and for leading the Government Procurement Service.

Notes: A practical manual of the considerations that a public sector client should make in choosing a procurement method.

Green, S D (1989) Tendering: optimization and rationality. *Construction Management and Economics*, 7(1), 52-63.
Abstract: There have been various techniques published for optimizing the net present value of tenders by the use of discounted cash-flow theory and linear programming. These approaches to tendering appear to have been largely ignored by the industry. This paper utilizes six case studies of tendering practice in order to establish reasons for this apparent disregard. Tendering is demonstrated to be a market oriented function with many subjective judgments being made regarding a firm's environment. Detailed consideration of "internal" factors such as cash-flow is therefore judged to be unjustified. Systems theory is then drawn upon and applied to the separate processes of estimating and tendering. Estimating is seen as taking place in a relatively sheltered environment and as such operates as a relatively closed system. Tendering, however, takes place in a changing and dynamic environment and as such must operate as a relatively open system. The use of sophisticated models to optimize the value of tenders is then identified as being dependent upon the assumption of rationality, which is justified in the case of a relatively closed system (i.e. estimating), but not for a relatively open system (i.e. tendering).
Notes: A good analysis of loading of items in the bill in estimating – front end, individual or back end. Not, however of relevance to the actual costs of the process.

Greenwood, D J (2000) *Power and proximity: a study of sub-contract formation in the UK building industry.* Unpublished PhD Thesis, Department of Construction Management & Engineering, University of Reading.
Abstract: The relationship between main contractors and sub-contractors is a major feature of construction activity. The importance of this relationship has been recognized, but its complexity has not. Attention has been focused exclusively on one dimension or another: either the parties' relative power or, more fashionably, their proximity. This leads to apparent anomalies: for example, where a small sub-contractor exerts an unexpected amount of power over a much larger main contractor; or a so-called "partnership" actually conceals the power of a dominant party. A new classification is proposed here, based on both dimensions. This power-proximity typology of relations is supported by legal, economic, and organizational literature, in which reference can be found to both dimensions and to the factors that affect them. A model of sub-contract formation was developed from this literature informed by the content of five in-depth interviews with practitioners. The model was examined and refined by generating hypotheses and testing them using empirical data from a survey of 88 transactions. The results clearly confirm the appropriateness of a typology of relationships based upon the two dimensions. In general, sub-contractors have less contractual power than main contractors, though some enjoy relatively higher levels than others. These were shown to relate to size, interdependence, contractual awareness and the availability of alternatives, with designers, piling specialists and lift installers exhibiting the highest levels. The most significant influence on proximity was the parties' past relationship, though contractors were less inclined to be close to sub-contractors that they perceived to be powerful or contractual. The conceptual map of relations provided by the study could prompt further research in a number of areas. Controlled replication of the study could result in refinement or development of the model itself. Alternatively attention may be directed towards its constituent elements. The concept of contractual power and the

indicators that have been chosen to measure it are potentially useful tools for analyzing contract formation. In terms of construction sub-contracts, attention should be directed from payment towards retention and damages terms as indicators of relative contractual power. Finally, in an era where there is currently a great deal of uncritical comment in the industry about the trend towards close and equitable relationships, a deeper understanding of the constructs that are associated with proximity, such as trust, mutuality, and reciprocity would be most valuable.
Notes: Presents a typology of relationships between main contractors and sub-contractors and a model of sub-contract formation that accounts for the relative power between the parties and their relative "proximity", in terms of the potential for cooperative relationships. The work provides a thorough explanation and analysis of contractual relationships at the sub-contract level but has no information of the costs of entering the various types of relationship described.

—— **(2001)** Sub-contract procurement: are relationships changing? *Construction Management and Economics*, **19**(1), 5-7.
Abstract: Recent publicity shows a shift in the attitude of main contractors to sub-contract procurement in the UK. However, a survey of the specialist contractors' sector shows that this impression should be approached with caution: the typical contractor/sub-contractor relationship is still traditional, cost-driven, and potentially adversarial. Nevertheless, the co-existence of the two approaches is consistent with the institutional theory of organizational strategy.
Notes: Provides empirical evidence on the extent of collaborative working at the contractor-sub-contractor level, and discusses the reasons for this as well as the persistence of more traditional "competitive" relationships.

Griffith, A, Knight, A and King, A (2003) *Best practice tendering for design and build projects*. London: Thomas Telford Publishing.
Abstract: The book is based on the findings of an EPSRC project about Design and Build (D&B) tender evaluation. The increased popularity of D&B offers the advantages of single-point responsibility, shorter project times and buildability. It also has the potential to allow more efficient supply chain solutions. The term covers a range of different procurement options with different risk distributions. The shortage of information about best practice techniques of bidding D&B and about proper tender evaluation leads to inefficiencies that puts upward pressure on prices in the industry. The conclusion is that more efficient use of suitable tender methodologies can reduce the time, resources and costs of the D&B tendering process.
Notes: The authors note that tender costs in D&B have always been considered higher than in traditional procurement (pp 100-102). A US-based contractor, interviewed as part of the study, claimed that on a $200m Federal Government project there were **wasted** tendering costs averaging $500,000 for each of the eleven tenderers. A particular problem is "scheme development" is required as part of a bid. One research respondent estimated these at $1m for a recently tendered scheme (value of the total bid not stated).

Griffiths, F H (1992) Bidding strategy: winning over key competitors. *Journal of Construction Engineering and Management, ASCE*, **118**(1), 151-65.
Abstract: This paper presents a method by which the probability of winning in the competitive bidding problem can be improved by obtaining additional information

concerning a key competitor. The competitive bidding problem as suggested by Friedman consists of describing the bidding of each competitor by developing a probability density function representing the distribution of the ratio of a competitor's bid to "our cost" estimate. This static distribution for a key competitor is improved by maintaining a volume-time function on competitor k. A function relating the mark-up of competitor k's static distribution function, thereby improving the calculation for the probability of winning over competitor k with the probability of winning over other competitors is also presented.
Notes: The focus of this paper is on bid mark-up decisions, rather than the cost of bidding, or any other transaction cost aspect of procurement.

Grimsey, D and Graham, R (1997) PFI in the NHS (private finance initiative in the UK National Health Service). *Engineering, Construction and Architectural Management*, **4**(3), 215-31.
Abstract: The National Health Service (NHS) hospitals in Britain are currently in a state of decay following many years of underinvestment in the estate. The NHS urgently requires billions of pounds of investment ranging from total hospital new builds to small refurbishment of existing facilities. The previous Conservative government put forward the Private Finance Initiative (PFI) as the procurement mechanism to address this problem. The new Labour government currently appear to be committing themselves to the same approach. PFI project sponsors have spent upwards of £30m bidding for around 30 major PFI schemes. Despite this, by the time of the UK election in May 1997, not one scheme had reached financial close and many sponsors were expressing their disillusionment with the process. Unlike PFI on other Government infrastructure and service schemes, each PFI hospital is tendered by a separate Trust with their own limited budgets. Many Trusts have demanded schemes without realizing that they cannot afford them and whilst these schemes may work out cheaper than publicly financed hospitals over 30 years or more, charges are higher in the early years. This is primarily due to the market for loans, the conditions attached to these loans in terms of repayment periods and cover ratios, and the requirement of the sponsors to generate a reasonable return on their investment. This paper discusses the major issues and analyses some of the technical financial problems surrounding the PFI in the NHS. The authors draw on practical experience of financial structuring and modelling hospital projects to build a generic model to analyse NHS PFI economics.
Notes: The authors mention that, at the time of writing, PFI sponsors had spent more than £30m on bidding for approx 30 schemes. They also mention, without substantiation, that "traditional arrangements primarily relied on standardized contract forms that allowed for the swift award of contracts at the expense of costly dispute resolution later in the process' (p. 218). Another interesting but unsubstantiated statement is "[t]here has been an underestimation by all parties of the length of time to negotiate project agreements" (p. 221). On page 221, the authors reveal that the source of their information is their activities as project finance advisers to sponsors of NHS PFI projects. The analytical part of the paper is the financial analysis that should take place in order to inform the investment decision. The unhelpful abstract of the paper is, in fact, an introduction as most of it does not appear anywhere else in the paper.

Grogan, T (1992) Low bids rise hidden costs. *Engineering News Record*, **228**(13), March 30, 30-1.
Abstract: The desperation bidding, low pricing by companies in the building process, in the US, as perceived by some industry sources, is having an effect on construction. While

owners are happy with these lower bids, it is likely there are however hidden costs to such low bidding, such as costs arising from contract changes, litigation, claims and contractor defaults. As a result, some owners are increasing the amount of funds available for bid contingencies. Steel prices over the past few years are compared to show their reeling prices. However, the lower prices of some materials, such as steel, could be the result of new modern mills that can operate at these lower prices.
Notes: The article is dated by particular market circumstances (those prevailing in early 1990s) but nevertheless there is an interesting assertion (unquantified) that low bids can result in (transaction) costs elsewhere.

Gruneberg, S L and Ive, G (2000) *The economics of the modern construction firm.* London: Palgrave.
Abstract: This is a book about the diverse range of firms that are found in the construction sector, about their decision-making and the economic environments in which they operate. The authors aim to provide a new and coherent perspective on these firms and the choices they have to make – both for their managers and for all who have a stake in these firms and their industry. It offers some new and important perspectives for research and teaching about construction firms.
Notes: Good general background work on the economics of the firm in a construction industry context. Includes discussions of transaction cost theory.

Guest, P (1997) Tender mercies. *Property Week, Careers Supplement,* Feb 1997, 14-15.
Abstract: Designed to offer private firms a share in local authority tenders, compulsory competitive tendering raised fears of job losses and a fall in new recruits to the public sector, however the reality is rather different.
Notes: This article is about the continuing, indeed rising, number of surveyors employed directly by local authorities in spite of CCT which should increase the involvement of the private sector.

Hameri, A-P and Nordberg, M (1999) Tendering and contracting of new, emerging technologies. *Technovation*, **19**(8), 457-65.
Abstract: This paper describes a practical model and procedure on how to tender products based on new, emerging technologies. This situation prevails in large-scale scientific projects and in other, mostly publicly funded, efforts to construct systems of significant technological scale. The paper proposes a double-blind communication and tendering procedure for establishing the means, partners and economic framework for the projects that finally commit to accomplish the task. This procedure sets a dedicated technological broker in to central position of the information exchange between the buyer and supplier. The presented procedure reduces any political interference, ensures more accurate cost estimates and helps the project to obtain better financial basis to kick-off. When in operation, the procedure can be used for determining the costs and most prominent technological trajectories for a proposed solution. The approach presented paves the ground towards the emergence of virtual markets for products yet-to-be-engineered.
Notes: Not set in the construction sector, but with some interesting, if futuristic, possibilities.

Hampson, K and Kwok, T (1997) Strategic alliances in building construction: a tender evaluation tool for the public sector. *Journal of Construction Procurement*, **3**(1), 28-41.
Abstract: Building construction is a highly competitive and risky business. This competitiveness is compounded where conflicting objectives amongst contracting and sub-contracting firms set the stage for an adversarial and potentially destructive business relationship. Clients, especially those from the public sector, need broader tender evaluation criteria to complement the traditional focus on bid price. There is also a need for change in the construction industry-not only to a more cooperative approach between the constructing parties – but also from a confrontationist attitude to a more harmonious relationship between all stakeholders in providing constructed facilities. A strategic alliance is a cooperative relationship between two or more organizations that forms part of their overall strategies, and contributes to achieving their major goals and objectives. Strategic alliances in building construction may provide a useful tool to assist public sector construction managers evaluate tenders and concurrently encourage more cooperative relationships amongst construction stakeholders. An overview of the Australian building construction industry is followed by a review of the existing strategic alliance literature and an analysis framework, comprising six attributes of strategic alliances for application to a construction organization: trust, commitment, interdependence, cooperation, communication, and joint problem solving. These attributes are currently being used to collect data from 70 building construction firms in Queensland to assess their respective levels of strategic alliance. Given the trend towards broader indicators of construction firm performance, these attributes are proposed as a tool for use in the tender evaluation process for public works.
Notes: Some relevance in terms of general background information on strategic alliances.

Harding, A, Lowe, D, Hickson, A, Emsley, M and Duff, R (1999) Implementation of a neural network model for the comparison of the cost of different procurement approaches. *In:* Hughes, W P (Ed), *15th Annual ARCOM Conference*, September 15-17, Liverpool John Moores University, UK. Association of Researchers in Construction and Management, Vol. 2, 763-72.
Abstract: The choice of procurement system for a building project is a very significant one. However, there is currently very little comparative cost data to inform the selection of a procurement system, especially the total cost to the client. This paper reports on a model that will make such a comparison possible. This model, which is currently under development at UMIST, is designed to consider 39 project variables, including the choice of procurement system, and estimate the final cost of the project using a neural network. The suitability of a neural network to model this problem has already been established by a pilot study. The advantage of using this type of model is that it permits comparisons between the different procurement methods to be made within the context of that particular project, rather than within projects as a whole. The classification and representation of the different variables to be considered within this model are discussed. In addition to this, the implementation of factor/cluster analysis to reduce the number of variables required by the neural network, and hence increase its accuracy, is explained. Furthermore, the level of confidence of the model and its implications for the implementation of a "What if?" analysis are discussed. This analysis would allow the client to assess how changing certain variables, including procurement, might affect the final cost of the project.
Notes: This is of relevance to the current project, though its concern is to produce a prototype model that predicts the effect of procurement method on *total* project cost, without distinguishing the transactional costs of the procurement process itself.

—— **(2000)** The cost of procurement: a neural network approach. *In:* Gudnason, G (Ed), *CIB W78 Construction Information Technology*, June 2000, Reykjavik, Iceland. CIB.
Abstract: Existing research that has attempted to determine differences between the costs of following different procurement routes has consistently aimed to determine a single blanket figure, such as "design and build is 15% cheaper than traditional". No attempt has yet been made to provide a difference which is project specific. Furthermore, no previous research has determined the total cost to the client using any objective method. The lack of data defining client costs and the absence of suitable modelling techniques have prevented such an objective evaluation being made. These factors prompted research into the cost of procurement at UMIST. This research has required the collection of a substantial database of the total cost, to a client, of past projects, and the subsequent creation of a neural network model of these costs. Results of the first phase of development of this model are presented, including regression analysis, preliminary neural network models and sensitivity analysis. An assessment of how these results will inform future development of the model is also made.
Notes: Refers to the same research project cited in the previous entry and is thus of some relevance to the current project.

Harrison, R S (1987) Managing the estimating function. *In:* Harlow, P (Ed) *Technical Information Service*, No. 75, Ascot: Chartered Institute of Building.
Abstract: The estimating process is a key function for a contractor's organization but the way in which it is run and its consequent effectiveness may vary greatly between firms. Careful and considered management of estimating is likely to produce a more efficient and effective result than the, often adopted, instinctive approach. The objective of this paper is to set out some of the main areas requiring consideration and the choices available in seeking the most suitable methods, structure and priorities for an estimating organization.
Notes: From an experienced estimator and provider of software that supports estimators in their work. Interestingly, he states that although it is a fundamental principle, obtaining work is not the main objective of the estimating process. On the costs of estimating, the author points out that the activity needs time, judgement and knowledge based on experience. Increased accuracy costs more, and the cost rises more rapidly than the increase in accuracy. Skilled estimators are scarce and expensive. This is all that is said about the cost of estimating. The remainder of the paper is concerned with providing advice to estimators, in a similar vein.

—— **(1993)** The transfer of information between estimating and other functions in a contractor's organization – or the case for going round in circles. *In: Construction Papers*, No. 19, Ascot, 1-8.
Abstract: This paper starts with some general thoughts on information and moves on to look at its existence at various stages of the construction process, who requires it and how it is provided. Finally, some alternative views of information flow are described.
Notes: Interesting background information on the flow of knowledge involved in tendering.

Hartman, F T (1994) Reducing or eliminating construction claims: a new contracting process (NCP). *Project Management Journal*, **25**(3), 25-31.
Abstract: Competitiveness can be a useful indicator of the long-term health of an industry. Whereas competitiveness has been studied at the corporate and national levels, its usefulness at the industry level has not been explored. The research objective is to develop a model to

evaluate competitiveness at the industry level. Motivated by persistent problems of the Canadian construction industry, an attempt is made to quantify its international competitiveness. The industry is compared with its counterparts in Japan and the United States. Competitiveness is defined, the different dimensions of competitiveness are illustrated, and the need to quantify competitiveness is discussed. A multi-criteria hierarchical model was developed and tested using both statistical and survey data. Salient findings of the research and conclusions are presented.
Notes: Construction claims represent one of the elements of transaction costs that the present study has identified. This is therefore of contextual relevance to the current project.

Herbert, C P and Biggart, P T (1993) Kingsford Smith airport, Sydney: planning and tendering the new parallel runway. *Proceedings of the Institution of Civil Engineers-Civil Engineering*, **97**(4), 182-9.
Abstract: This paper examines the present limitations of runway capacity for commercial aircraft in the Sydney region and the reasons for deciding to build a third runway at the existing Kingsford Smith airport. The location and preliminary design of the preferred alternative are discussed, as is the extensive environmental examination which resulted in conditional approval to proceed. Alternative project delivery systems are outlined, resulting in the decision to appoint a managing contractor to undertake the detailed design and construction. Lastly, the tender process is explained, showing how tenders were assessed and the reasons for choosing the successful contractor for this major project.
Notes: The tender process, tender assessment and contractor selection processes are thoroughly explained, confirming the complexity of tendering and the difficulty of measuring various factors.

Hewes, M (2001) 50 top clients. *Building, Building Directory Supplement*, December, 1-44.
Abstract: This guide brings together key information on 50 of the construction industry's major clients. Between them, these companies spend close to £24bn in the UK, which translates into at least £16bn on construction work. An analysis of future capital investment plans reveals that, of the top 50, about 55% intend to raise expenditure, 30% are to keep it level, and 15% will probably reduce it. However given the fast-changing economic environment, some of these plans will alter. In terms of procurement, close to 55% of the companies still use traditionally-based contracts in one form or another, whereas 45% use negotiated contracts, whether through partnering, framework, or alliancing. Almost 30% of clients use PFI contracts in some form. Many of those who do not use negotiated contracts are considering doing so. Design-and-build remains common, especially among developers, many of which use it alongside other forms of contract. Finally, most clients use term contracts for their maintenance work. This directory is the most comprehensive client information Building has ever published.
Notes: A source of information on the methods of procurement employed by clients in the industry.

Hillebrandt, P M and Cannon, J (Eds) (1989) *The management of construction firms: aspects of theory*. Basingstoke: Macmillan.
Abstract: Can the construction industry cope with the challenges of the future? Is it fitter and leaner or thinner and weaker? This book presents a challenging analysis of the state of large construction companies. It focuses on the changes in their environment and behaviour from the boom conditions of the late 1980s, the decline in the firms' traditional markets and

their attempts to develop others, the disastrous financial experiences of the early 1990s, and the changes in strategies and structures and in the management of the firms. It is based on published data and interviews with senior executives of twenty major companies.
Notes: A background research-based text in construction industry economics.

Hillebrandt, P M and Hughes, W P (2000) What are the costs of procurement and who bears them? *In:* Ngowi, A B and Ssegawa, J (Eds), *Challenges facing the construction industry in developing countries, 2nd International Conference of the CIB Task Group 29*, 15-17 November 2000, Gabarone, Bostwana. Botswana National Construction Council (BONCIC, Faculty of Engineering and Technology, University of Botswana, Council for Research and Innovation in Building (CIB), Vol. 1, 415-20.
Abstract: The costs of procurement are transaction costs, which are separate from the direct costs of a project. In this paper discussion is concentrated on costs of tendering. Types of cost, including money costs and opportunity costs, short-term and long-term costs, private and social costs are defined and examined in relation to various types of product and methods of procurement. The costs of the contractor and of the client are considered and tentative conclusions drawn as to who bears these costs in the short-run and in the long run. They may fall on the parties to the process for the particular project, on other contractors and clients or on society as a whole.
Notes: An intermediate output from the research into the costs of procurement.

Hines, P and Rich, N (1997) The seven value stream mapping tools. *International Journal of Operations and Production Management*, **17**(1), 46-64.
Abstract: Develops a new value stream or supply-chain mapping typology. This seven-map typology is based on the different wastes inherent in value streams. The use of the various tools, either singularly or in combination, is therefore driven by the types of waste to be removed. The tools themselves are drawn from a range of existing functional ghettos such as logistics, operations management and engineering. Maintains that two of the seven tools can be regarded as completely new. This cross-functional approach means that the choice of tools to be used can be made from outside of traditional departmental boundaries, affording researchers and companies the opportunity to use the most appropriate tools rather than merely those that are well-known in their function. Describes each tool briefly and gives a simple mechanism for choosing which is most appropriate to contingent situations.
Notes: Not strictly relevant to the costs of procurement but of interest in the context of supply chains and their mapping.

Hinze, J and Tracey, A (1994) The contractor-sub-contractor relationship: the sub-contractor's view. *Journal of Construction Engineering and Management, ASCE*, **120**, 274-87.
Abstract: In the construction of most projects, a significant role is played by specialty contractors, also commonly referred to as sub-contractors. Despite the importance of sub-contractors, little is publicized about the actual process by which sub-contracts are initiated, or how award arrangements are made. Therefore, an exploratory study was conducted that focused on this subject. Information was obtained on bidding practices, sub-contracting arrangements, administrative practices, payment procedures, and project closeout. The results provide information on various methods used by general contractors to place sub-contractors at risk. Bid shopping appears to be a continuing practice in the construction industry, with little recourse for sub-contractors. Sub-contractors are often contractually

required to assume risks that they would not otherwise assume. They are often required to assume all the obligations as stipulated in the contract between the owner and the contractor, but are not afforded the opportunity to examine it. Payment problems continue for sub-contractors, with the practice essentially accepted by sub-contractors as being a part of doing business. Suggestions are offered for improving the relationship between sub-contractors and general contractors.
Notes: A viewpoint on the important contractor-sub-contractor relationship.

Holt, G D, Olomolaiye, P O and Harris, F C (1993) A conceptual alternative to current tendering practice. *Building Research and Information*, **21**(3), 167-72.
Abstract: Alternative quantitative selection techniques developed into an expert system.
Notes: A paper on contractor selection. Argues against subjectivity and price criteria. The paper presents a history of the development of selective (as opposed to open) tendering. The authors suggest a contractor selection model based on three (P1, P2, P3) scores: P1 is pre-qualification (with 5 factors, 18 variables), P2 is mid-tender (with 2 factors, 9 variables), and P3 is at tender stage. The one factor at bid stage is a datum based on the client's own "in-house" estimate.

—— **(1996)** Tendering procedures, contractual arrangements and Latham: the contractors' view. *Engineering, Construction and Architectural Management*, **3**(1 & 2), 97-115.
Abstract: The procedural and administrative aspects of UK tendering have remained largely unaltered for decades but this may soon change in light of the recent review of the construction sector, headed by Sir Michael Latham. This paper presents findings of a nationwide survey of UK construction contractors assessing their opinion of the Latham procurement recommendations, along with their opinion of the authors' proposals for alternative selection procedure. Contractor usage/opinion of current tendering methods, tendering documentation and contractual arrangements are also identified. Analysis techniques primarily involve the derivation of contractor preference, agreement and importance indices (Pr, Ag, and Im respectively). Results show that clients are attempting to cut costs by increased use of open tendering coupled with plan and specification tender documentation, but that savings are offset by clients ultimately paying for contractor's costs in achieving contract award for only one in five bids. Generally, contractors are in tune with ideas of the Latham review and characteristics pertaining to the HOLT (Highlight Optimum Legitimate Tender) selection technique.
Notes: The coyness of some people about the costs of tendering are revealed here by the phrase "...savings are offset by clients ultimately paying for contractors' costs associated with their achieving contract award for only one in five cases". The paper is based on a survey sample of 41 companies. The paper identifies some very important questions, but their phrasing, coupled with the tiny sample and strong reliance on anecdotes, renders the finding somewhat unreliable. There is some important potential here that is not realized. The appendix contains some useful direct quotations from the contractors surveyed.

Hook, M (1987) Fee tendering: RIAS code of procedure for fee tendering. *Architects Journal*, **185**(18), 77.
Abstract: The lottery of fee competition, now in many cases a necessity, has very few ground rules. Michael Hook looks at the code of procedure for fee tendering published by the Royal Incorporation of Architects in Scotland and assesses its attempts to assist both client and architect.

Notes: This article looks and assesses the RIAS code of procedure for fee tendering, which fails to address the complex reality of composite fee bidding. There is little in this article relevant to the current project.

Hoxley, M (2000) Are competitive fee tendering and construction professional service quality mutually exclusive? *Construction Management and Economics*, **18**(5), 599-605.
Abstract: It is a little more than 15 years since the associations representing construction professionals in the UK surrendered to government pressure and abolished mandatory fee scales, predicting as they did so that inevitably abolition would lead to a decline in the standard of service provided to clients. Initially the abolition of fee scales had little impact on fee levels – in the UK economic and property boom of the late 1980s demand from clients in all sectors was high and fee levels remained at, or close to, pre-abolition levels. However, in the recession that followed, fee levels fell to unprecedented low levels, causing many commentators to be concerned that the quality of service provided to clients would fall. The main aim of this research is to establish whether clients' perceptions of service quality have declined as a result of lower fee scales. Following a literature search five hypotheses are presented, namely, that clients' perceptions of service quality are: lower for fee tendered commissions; lower when the fee bid is particularly competitive; higher when the service is adequately specified by the client; higher when care has been taken with pre-selection of tenderers; and higher when adequate weighting to ability is given in the final selection process. The hypotheses have been tested by collecting data from 244 clients who anonymously assessed consultants, 60% of whom were chartered surveyors (just over half of these were quantity surveyors). Over half of the consultants were appointed by competitive fee tender, and although the service quality scores were lower for these consultants this result was not statistically significant. Therefore, the main hypothesis is not supported by the data but the fourth and fifth hypotheses are both supported by the study.
Notes: The question posed in the title was investigated by sending a questionnaire to 500 clients (over half local authorities), 244 of whom responded. They were asked to rate 26 statements about consultants they had employed. It was found that the abolition of mandatory fees has not reduced the quality of service. SERVQUAL is used to measure quality.

Hughes, M (1997) Tender evaluation in CCT: developing quality price models. *Proceedings of the Institution of Civil Engineers-Municipal Engineer*, **121**(2), 108-11.
Abstract: This paper looks at the merits of evaluating "quality" in compulsory competitive tendering (CCT) and the various methods by which this may be achieved.
Notes: This paper assesses ways in which local authorities can include quality assessment in the award of tenders as well as prices and still clearly be objective in their choice of contractor. The paper links with other papers, notably an investigation of tendering for co-operation-municipality contractors in the Netherlands, by Dorée (1996).

Hughes, W P, Hillebrandt, P M, Greenwood, D and Kwawu, W (2002) Developing a system for assessing the costs associated with different procurement routes in the construction industry. *In:* Uwakweh, B O and Minkarah, I A (Eds), *The 10th International Symposium of the W65 Commission on Organization and Management of Construction*, 9-13 September 2002, University of Cincinnati, USA. CRC Press, New York, Vol. 2, 826-37.
Abstract: In developing techniques for monitoring the costs associated with different procurement routes, the central task is disentangling the various project costs incurred by

organizations taking part in construction projects. While all firms are familiar with the need to analyse their own costs, it is unusual to apply the same kind of analysis to projects. The purpose of this research is to examine the claims that new ways of working such as strategic alliancing and partnering bring positive business benefits. This requires that costs associated with marketing, estimating, pricing, negotiation of terms, monitoring of performance and enforcement of contract are collected for a cross-section of projects under differing arrangements, and from those in the supply chain from clients to consultants, contractors, sub-contractors and suppliers. Collaboration with industrial partners forms the basis for developing a research instrument, based on time sheets, which will be relevant for all those taking part in the work. The signs are that costs associated with tendering are highly variable, 1-15%, depending upon what precisely is taken into account. The research to date reveals that there are mechanisms for measuring the costs of transactions and these will generate useful data for subsequent analysis.
Notes: An interim report from the research into the costs of tendering that led to this publication.

Hughes, W P, Hillebrandt, P M, Lingard, H and Greenwood, D (2001) The impact of market and supply configurations on the costs of tendering in the construction industry. *In:* Duncan, J (Ed), *CIB World Building Congress*, April 2001, Wellington, New Zealand. CIB.
Abstract: The cost of tendering in the construction industry is widely suspected to be excessive, but there is little robust empirical evidence to demonstrate this. It also seems that innovative working practices may reduce the costs of undertaking construction projects and the consequent improvement in relationships should increase overall value for money. The aim of this proposed research project is to develop mechanisms for measuring the true costs of tendering based upon extensive in-house data collection undertaken in a range of different construction firms. The output from this research will enable all participants in the construction process to make better decisions about how to select members of the team and identify the price and scope of their obligations.
Notes: An interim report from the research into the costs of tendering that led to this publication

Hughes, W P and Murdoch, J R (2001) *Roles in construction projects: analysis and terminology*. London: Construction Industry Publications.
Abstract: Standard form contracts are typically developed through a negotiated consensus, unless they are proffered by one specific interest group. Previously published plans of work and other descriptions of the processes in construction projects tend to focus on operational issues, or they tend to be prepared from the point of view of one or other of the dominant interest groups. Legal practice in the UK permits those who draft contracts to define their terms as they choose. There are no definitive rulings from the courts that give an indication as to the detailed responsibilities of project participants. The science of terminology offers useful guidance for discovering and describing terms and their meanings in their practical context, but has never been used for defining terms for responsibilities of participants in the construction project management process. Organizational analysis enables the management task to be de-constructed into its elemental parts in order that effective organizational structures can be developed. Organizational mapping offers a useful technique for reducing text-based descriptions of project management roles and responsibilities to a comparable basis. Research was carried out by means of a desk study, detailed analysis of nine plans of work and focus groups representing all aspects of the construction industry. No published

plan of work offers definitive guidance. There is an enormous amount of variety in the way that terms are used for identifying responsibilities of project participants. A catalogue of concepts and terms has been compiled and indexed to enable those who draft contracts to choose the most appropriate titles for project participants. The purpose of this terminology is to enable the selection and justification of appropriate terms in order to help define roles.
Notes: A useful de-construction of the management tasks that are pre-supposed in standard form contracts. This provides a more detailed account than most about the steps through which construction projects typically pass.

Ibbs, W C and Ashley, D B (1987) Impact of various construction contract clauses. *Journal of Construction Engineering and Management, ASCE,* **113**(3), 501-21.
Abstract: Construction contract types and general conditions clauses have a major influence on the likelihood and degree of project success. A comprehensive empirical study was conducted for 36 very large capital construction projects completed within the past three years to assess such impacts. Two general contract types (cost reimbursable and fixed price) and 96 clauses were carefully analysed for impacts on six measures of project performance: cost, schedule, quality, safety, and owner and contractor satisfaction. The clauses bearing most heavily on project performance are identified, and key elements of those clauses that are most crucial to project success are discussed in detail. Other contract administration concepts and incentive provisions are factually and thoroughly discussed.
Notes: An important argument showing that with different "contract types" a project's transaction costs are transferable to different stages of the construction process.

Institute of Civil Engineers, Association of Consulting Engineers and Federation of Civil Engineering Contractors (1983) *Guidance on the preparation, submission and consideration of tenders for civil engineering contracts: recommended for use in the United Kingdom.* London: Institution of Civil Engineers, Association of Consulting Engineers and Federation of Civil Engineering Contractors.
Notes: An ICE version of the NJCC Code of Tendering (superseded by the CIB Codes, *q.v. infra*).

Jackson-Robbins, A (1999) Selecting contractors by value. *CIRIA Special Publication 150,* London: Construction Industry Research and Information Association.
Abstract: Explores ways in which clients may choose contractors according to the work they do and the value that is invested.
Notes: A practical manual to support the adoption of methods of selection that are not entirely price-based.

Jennings, P and Holt, G D (1998) Pre-qualification and multi-criteria selection: a measure of contractor's opinions. *Construction Management and Economics,* **16**(6), 651-60.
Abstract: The procurement of a construction contractor normally involves some form of pre-qualification. The better pre-qualification regimes adopt a structured multi-criteria approach (i.e. contractor evaluation based on a multiple of factors over and above cost). This research solicits contractors' viewpoints on pre-qualification, in contrast to earlier works which have tended to present clients' perspectives. Contractors are dissatisfied with the frequency and adequacy of current pre-qualification regimes. An investigation of the mutual benefits of multi-criteria selection leads to the suggestion that such benefits could be

better capitalized upon. Contractors' perceived level of importance (with respect to selection criteria considered by client during the selection process) are evaluated and show that, in line with earlier findings, "cost" is still the predominant selection factor, followed by "contractor experience" and "company reputation". A comparison between contractors' rankings of the selection criteria with similar rankings derived from an earlier survey of clients, finds significant correlation, indicating that contractors agree with clients' importance levels of multi-criteria selection factors.

Notes: This is an interesting paper looking at the contractors' views of selection procedures. It is based on a questionnaire sent to 80 contractors and completed by 37 of mixed sizes. Some replies relate to pre-qualification for a standing list; others to qualification for a particular project. The criteria include, for example, price, company experience, financial standing, experience and qualifications of personnel. There is considerable agreement of these results with those of clients found in other studies as to the relevance of various criteria. As gathering information to give to a client and maintaining it has a cost, this raises questions about who does this and at what cost. Similarly, how much time does the client spend on assessment of the data?

Jones, M (1992) Will CCT take architecture back to the sixties? *Architects Journal*, **196**(24), 19-21.

Abstract: Colin Stansfield-Smith fears damage from compulsory competitive tendering; lessons learned from privatization of Surrey county architects.

Notes: With the government's legislation on compulsory competitive tendering for architectural services, many public architects feared multiple damage of a system that takes no account of the complex nature of the local authority client. For example, some aspects of CCT imply that the value assigned to a feasibility study is somehow equated with that of a redecoration specification. In the long run price will be the lowest common denominator and cheap the overriding word and everybody making the cheapest contribution. CCT would disrupt highly beneficial symbiotic relationships as sometimes one wants to negotiate on price scale, on design as to the skills of a particular designer or quantity surveyor.

Kagioglou, M, Cooper, R, Aouad, G and Sexton, M (2000) Rethinking construction: the generic design and construction process protocol. *Engineering, Construction and Architectural Management*, **7**(2), 141–53.

Abstract: The complexity of construction projects and the fragmentation of the construction industry undertaking those projects has effectively resulted in linear, uncoordinated and highly variable project processes in the UK construction sector. Research undertaken at the University of Salford resulted in the development of an improved project process, the Process Protocol, which considers the whole lifecycle of a construction project whilst integrating its participants under a common framework. The Process Protocol identifies the various phases of a construction project with particular emphasis on what is described in the manufacturing industry as the fuzzy front end. The participants in the process are described in terms of the activities that need to be undertaken in order to achieve a successful project and process execution. In addition, the decision-making mechanisms, from a client perspective, are illustrated and the foundations for a learning organization/industry are facilitated within a consistent Process Protocol.

Notes: An interesting alternative to traditional approaches for portraying the process of construction project procurement. It does not provide sufficient detail for analysing costs.

Kashiwagi, D T, Halmrast, C T and Tisthammer, T (1996) "Intelligent" procurement of construction systems. *Journal of Construction Procurement*, **2**(1), 56-65.
Abstract: The methodology reshapes the construction environment, changes roles and partnerships, and provides facility owners with a new method of communicating requirements to constructors based on performance information and "intelligent thinking." The new performance approach allows facility owners to communicate requirements in simple performance terms instead of material and construction standards. The methodology was developed using the industrial engineering concepts of continuous improvement, quality control, and intelligent thinking processes. The process has been refined and tested fifteen times from 1994-1996 on the procurement of roofing systems, janitorial services, and landscaping services.
Notes: Another of the PBSRG's papers supporting the minimization of management effort.

Kashiwagi, D T and Mayo, R E (2001) Best value procurement in construction using artificial intelligence. *Journal of Construction Procurement*, **7**(2), 42-59.
Abstract: There is a growing movement in the construction industry away from procurement through traditional low-bid methods toward performance-based and best-value procurement. Best value is achieved by contractors who optimize both price and performance. Until now there have been two major types of best value procurement: one-step and two-step. The major difficulty with both systems is finding an unbiased way to evaluate value vs. performance. This study explores best value as identified by artificial intelligence (AI). This paper covers a case study of the use of a performance-based procurement system, called Performance Information Procurement System (PIPS), to obtain best value on roof replacement and mechanical systems replacement projects in public schools in the state of Hawaii.
Notes: The introduction of the computerized PIPS system to minimize selection costs.

Kennedy, P, Morrison, A and Milne, D O (1997) Resolution of disputes arising from set-off clauses between main contractors and sub-contractors. *Construction Management and Economics*, **15**(6), 527-37.
Abstract: Set-off relates to the situation where a main contractor raises a counterclaim against a sub-contractor's claim or where an employer raises a counterclaim against the main contractor. The alternative terms cross claim, counterclaim, contra charge, compensation and retention are explained in the context of Scots law. Set-off in the construction industry in Scotland is then discussed within the contractual frameworks upon which main contractors are entitled to exercise such rights and how these conditions of contract have been formed over recent years. A study shows the extent of the use of amended and unamended forms of sub-contract and main contractors' own forms of sub-contract which imposed more onerous set-off conditions than the standard forms, the reasons given by main contractors for exercising their rights of set-off, the level of satisfaction amongst sub-contractors with the sums set-off against them, the means by which main contractors and sub-contractors settled disputed set-offs, and sub-contractors' reasons for accepting unsatisfactory instances of set-off. The research was undertaken using a questionnaire to a stratified sample of sub-contractors throughout central Scotland in 1995. Forty-seven sub-contractors responded to the questionnaire and 427 instances of set-off were recorded. The study indicated that, despite the considerable protection given to sub-contractors in the standard forms of sub-contract and in common law, they were prepared to settle set-offs with which they were dissatisfied without initiating contractual proceedings

which would have improved their situation. It would appear from this study that sub-contractors are reluctant to use their contractual entitlements either because of fears over the costs of disputing set-offs or because they fear that they will be denied opportunities to tender for work in the future. Until there is a culture shift in the industry, reliance on contractual conditions alone may be inadequate to meet the needs of sub-contractors.
Notes: One of the aspects of transaction cost that is of interest to the present study is that concerned with enforcement of the deal. The legal and contractual wrangling concerning set-off is a good example of how such costs arise.

Khosrowshahi, F (1999) Neural network model for contractors' pre-qualification for local authority projects. *Engineering, Construction and Architectural Management*, **6**(3), 315-28.
Abstract: The way in which clients or their consultants undertake to select firms to tender for a given project is a highly complex process and can be very problematic. This is also true for public authorities as, for them, "compulsory competitive tendering" is a relatively new concept. Despite its importance, contractors' pre-qualification is often based on heuristic techniques combining experience, judgement and intuition of the decision-makers. This, primarily, stems from the fact that pre-qualification is not an exact science. For any project, the right choice of the contractor is one of the most important decisions that the client has to make. Therefore, it is envisaged that the development of an effective decision model for contractor pre-qualification can yield significant benefits to the client. By implication, such a model can also be of considerable use to contractors: a model of this nature is an effective marketing tool for contractors to enhance their chances of success to obtain new work. To this end, this work offers a decision-support model that predicts whether or not a contractor should be selected for tendering projects. The focus is on local authorities because, in the absence of a viable universal selection system, there are significant variations in the way they conduct pre-qualification. The model is based on the use of artificial neural networks (ANN) and uses data relating to 42 local authorities (clients). Tests reveal that the model has a highly satisfactory predictive accuracy and that the ANN technique is a viable tool for the prediction of success or failure of the contractor to qualify to tender for local authority projects.
Notes: Focused on the pre-qualification stage of the tender process. Dated reference to CCT in Local Authorities even though the paper dates from the year CCT was abolished (1999). Offers decision support tool for contractor pre-qualification. Pre-qualification is an important issue for clients and contractors (and a cost to both). A typical pre-qualification process is illustrated (p.317). The cited works provide a good grounding in what could be involved in contractor pre-qualification, identifying 21 variables.

King, M and Mercer, A (1991) Distributions in competitive bidding. *Journal of the Operational Research Society*, **42**(2), 151-55.
Abstract: Analyses of historical tendering data for a company selling concrete farm buildings have called into question the usual assumption of normality or log-normality for the probability distribution. The desirability of modelling individual competitors' strategies is reinforced. Complex problems of parameter estimation and testing have been overcome by adopting a likelihood approach, with values examined for plausibility.
Notes: A theoretical paper dealing with the variability of bids.

Kodikara, G W and McCaffer, R (1993) Flow of estimating data in Sri Lankan building contractor organizations. *Construction Management and Economics*, **11**(5), 341-6.

Abstract: An estimator prices a bill of quantities, collects, generates and assembles data (estimating data) for the purpose of establishing the cost of constructing the project. The data generated could be used by the contractor's subsequent management functions, and the use of estimating data in the contractors' post-tender management is worthy of attention. Drawing information from ten case studies of the organization of Sri Lankan building contractors, this paper identifies the contractors' management functions. These flows highlight the substantial burden of re-work in the post-tender use of data. It is argued that the current format and presentation of estimating data in Sri Lanka are the major causes for such re-work. However, it was found that any revolutionary change to the conventional format would not be welcomed by the industry. Any new proposal should be developed within the limitation of acceptability to conventional practice. The recommendation is the "unit rate" is broken down to its cost components of materials, labour and plant. The breakdown of the unit rate would supply all the necessary data for direct use, thus reducing the re-work. Further research should be addressed to investigating the best format and structure of this breakdown.
Notes: More evidence about the normal flow of information (a costly item) in the preparation of contractors' tenders.

Kodikara, G W, Thorpe, A and McCaffer, R (1993) The use of bills of quantities in building contractor organizations. *Construction Management and Economics*, **11**(4), 261-9.
Abstract: The prime purpose of the Bill of Quantities (BQ) is to enable all contractors tendering for a contract to price on exactly the same information. Subsequent to this, it is widely used for post-tender work such as: material scheduling; construction planning; cost analysis; and cost planning. Due to the re-work involved in the post-tender use of the BQ, the extent of use of the BQ is important. The extent of use is defined as the direct use, after subtracting the re-work from the total use. This paper identifies contractors' current use of the BQ for post-tendering work based on eight case studies, and establishes the extent of use thus highlighting the re-working of the bill. By establishing the extent of use, the true picture of the direct use and the repetition work can be shown. The average extent of use of the BQ for post-tender work in the industry was found to be 50%. This 50% use of the BQ requires some form of re-working. This re-work needs to be reduced if improved post-tender use of estimating data is to achieved. Information stored in the BQ should be arranged in a directly useable way. It was found that, quantities, quantity units, and unit rates are the key elements of the BQ information that need to be presented in a more meaningful format if the amount of re-work is to be reduced.
Notes: The BQ is an interesting vehicle for manipulating the cost of tendering. A more detailed (hence more costly BQ) will probably produce a more economically derived set of tenders.

Kumaraswamy, M M and Dissanayaka, S M (1998) Linking procurement systems to project priorities. *Building Research and Information*, **26**(4), 223-38.
Abstract: The main objective of this work is to assess the impacts of various procurement variables on project performance, in comparison to the impacts of non-procurement related variables, such as project conditions and team characteristics. Since the impacts of specific procurement variables, such as contract type, have been investigated before, it was decided to first develop a comprehensive framework of procurement options. This framework is based on a holistic overview of procurement systems that included, for example, sub-systems of work packaging, and type of contract. A model was next developed to link the

procurement system variables to project outcomes. Additional variables such as the characteristics of the project and the project participants were incorporated into this model as "intervening" variables. A survey in Hong Kong tested the relative strengths of such procurement sub-systems and intervening variables, in terms of their influences on project outcomes. The interim results are described in this paper. This includes observations that cost and time overruns were not significantly influenced by the chosen intervening variables. Such observations led to the identification of particular needs to further probe the influences of team performance levels, as well as of procurement sub-systems. This evaluation of the other relationships in this proposed model will assist clients and their advisers to design more appropriate procurement systems that should be geared to their particular project priorities.
Notes: A comprehensive framework of procurement options is offered, together with the relative attractions of these options.

Kumaraswamy, M M and Dulaimi, M F (2001) Empowering innovative improvements through creative construction procurement. *Engineering, Construction and Architectural Management*, **8**(5/6), 325-34.
Abstract: The heightened state of flux in the construction industry in general, and construction procurement strategies in particular, provides welcome opportunities to inject innovative improvements. While some improvements are generated from within the construction industry itself, these evolve sluggishly along prolonged learning curves. These are compared with lessons to be learned and examples to be drawn from manufacturing in the development of a marketable product. A product development focus is thus advocated in re-integrating segregated groups and in empowering and inspiring the innovations that are needed to achieve the dramatic productivity gains now demanded from the construction industry.
Notes: The paper compares the responsiveness of the construction industry with that of manufacturing when presented with non-traditional procurement methods.

Laedre, O and Haugen, T I (2002) Target pricing in partnering projects: Examining the effect of integrated project teams and target pricing in three pilot projects. *In:* Lewis, J M (Ed), *CIB W92 Procurement systems*, January 2002, Trinidad. The Engineering Institute, University of the West Indies, St Augustine, Trinidad & Tobago, Vol. 1, 14.
Abstract: In Norway there is a trend in organizing the building process with focus on better integration of the different parties and use of new procurement methods. In their recent studies, the authors have evaluated three pilot construction projects, two small road projects and one railway crossing point, all involving a tunnel and a roadbed. One of the projects was classified as a research project and based on a negotiated contract, one contract was based on competitive bidding among pre-qualified contractors and one contract was made between two separate divisions within the same public agency. The goals in these projects have been to create better integration and co-operation between the clients, the external consultants and the contractor. This integration should be leading to a better result with respect to total costs and quality. The contracts between the public clients and contractors have been based on an agreed target price with incentives linked to the final costs. The main impression is that the project participants had a share of positive experiences that was predominant to the share of negative experiences. The participants chose better and more cost effective technical solutions during both the programming and production period, and they considered the partnering models as inspiring.

Notes: Case study research that investigates the outcomes in more integrated construction projects.

Lam, B C L (1998) Management and procurement of the Hong Kong Airport Core Programme. *Proceedings of the Institution of Civil Engineers-Civil Engineering*, **126**(SI1), 5-14.
Abstract: This paper describes the ten main elements of the US$ 20 billion Hong Kong Airport Core Programme, ranging from the 1248 ha airport platform at Chek Lap Kok to the dramatic series of bridges, viaducts and tunnels which make up the 34 km road and rail link to the Central business district. It discusses the formidable problems and constraints which faced the 35 000 construction workers and explains the complex management and control structures which enabled this huge project to be completed on time and within budget. Tendering procedures, conditions of contract and dispute resolution mechanisms are also reviewed.
Notes: An interesting study of the management and procurement of one of the world's largest projects of its time.

Lambert, D M, Cooper, M C and Pagh, J D (1998) Supply chain management: implementation issues and research opportunities. *International Journal of Logistics Management*, **9**(2), 1-19.
Abstract: In 1998, the Council of Logistics Management modified its definition of logistics to indicate that logistics is a subset of supply chain management and that the two terms are not synonymous. Now that this difference has been recognized by the premier logistics professional organization, the challenge is to determine how to successfully implement supply chain management. This paper concentrates on operationalizing the supply chain management framework suggested in a 1997 article. Case studies conducted at several companies and involving multiple members of supply chains are used to illustrate the concepts described.
Notes: A paper that lends theoretical status to the newly-emerged field of supply chain management. Because of its characteristics, construction is an industry in which supply chains are particularly important.

Latham, M (1993) *Trust and money: interim report of the joint government/industry review of procurement and contractual arrangements in the United Kingdom construction industry.* London: Department of the Environment.
Abstract: not available.
Notes: This is the interim report of Sir Michael Latham. The aim was to articulate the problems, for which solutions would be offered in the final report (Latham 1994). The lack of trust and obstructions in the flow of money were identified as chief amongst the contractual problems of the industry.

—— **(1994)** *Constructing the team: final report of the government/industry review of procurement and contractual arrangements in the UK construction industry.* London: HMSO.
Abstract: not available.
Notes: This report tackles the problems identified in the interim report (Latham 1993). Selection, tendering, contractual design and construction procedures are all targeted. Its

main objectives are to promote a team approach to all aspects, and the promotion of a high quality, high efficiency industry.

Levene, P, Jackson, N, Gray, R, Jensen, J, Massey, G, Moschini, S, West, R and Woodman, R (1995) *Construction procurement by government: an efficiency unit scrutiny (The Levene Report)*, London: Efficiency Unit Cabinet Office.
Abstract: Government expenditure on construction projects is important to both the taxpayer and the UK construction industry. Each year, Government bodies spend about £6 billion on works including courts, hospitals, laboratories, military installations, offices, prisons and roads. This report examined construction procurement across Government to find ways of improving its efficiency and making Government a "best practice client". In general, the industry was found lacking in customer focus, often highly adversarial, and too ready to use any excuse to pursue additional payments by the client (so-called "claims"); fragmented and even divided against itself, with poor communications and little mutual respect between contractor/constructors, specialist suppliers and the profession; under-capitalized and operating on low margins; and perhaps as a consequence of all these, slow to adopt modern technology or approaches to quality service delivery.
Notes: As key clients of the industry, Government bodies are far from blameless for the way in which those bodies relate to it. To get the industry to change its way, Government will have to change its own behaviour, practice and procedures. The issue is not one-sided. Five Action Points are proposed, explained and developed into 22 specific recommendations.

Ling, Y Y and Boo, J H S (2001) Improving the accuracy of approximate estimates of building projects. *Building Research and Information*, **29**(4), 312-18.
Abstract: This paper investigates the level of estimating accuracy in Singapore between 1992 and 1998. It compares the actual estimating accuracy against practitioners' expectations, and past studies. Forty-two sets of project data were obtained from six Singapore based quantity surveying firms. The estimating accuracy was found to be 1.09. This level of accuracy is below the expectation of 41 quantity surveyors (QS) who were surveyed. There is therefore a need to increase estimating accuracy further. Results also showed that there is no significant difference in estimating accuracy between this study and a similar, earlier study. The implication of this finding is that longitudinally, there is no improvement in estimating accuracy over time. Moreover, for future studies on estimating accuracy, researchers can rely on cost data sets that are relatively dated, and from one source. Thirteen possible methods to improve estimating accuracy were tested in the fieldwork. It was found that the more important methods relate to having good quality and sufficient quantity of design information to prepare estimates.
Notes: There is limited relevance to the present study by way of the relationship between increased accuracy and increased (tendering) costs.

Lingard, H, Hughes, W P and Chinyio, E A (1998) The impact of contractor selection method on transaction costs: a review. *Journal of Construction Procurement*, **4**(2), 89-102.
Abstract: The basic premise of transaction-cost theory is that the decision to outsource, rather than to undertake work in-house, is determined by the relative costs incurred in each of these forms of economic organization. In construction the "make or buy" decision invariably leads to a contract. Reducing the costs of entering into a contractual relationship (transaction costs) raises the value of production and is therefore desirable. Commonly applied methods of contractor selection may not minimize the costs of contracting.

Research evidence suggests that although competitive tendering typically results in the lowest bidder winning the contract this may not represent the lowest project cost after completion. Multi-parameter and quantitative models for contractor selection have been developed to identify the best (or least risky) among bidders. A major area in which research is still needed is in investigating the impact of different methods of contractor selection on the costs of entering into a contract and the decision to outsource.
Notes: The observation that "commonly applied methods of contractor selection may not minimize the costs of contracting" is a key starting point for the present study.

Lo, W, Chao, C H, Hadavi, A and Krizek, R J (1998) Contractor selection process for Taipei Mass Rapid Transit System. *Journal of Management in Engineering, ASCE*, **14**(3), 57-65.
Abstract: It is normal practice for developing countries to use a large industrial or technological project as a means to transfer foreign expertise and experience into the country, and the government of Taiwan has pursued this path in building the Taipei mass rapid transit system (MRTS). To satisfy the foregoing objective, the Department of Rapid Transit System (DORTS) set the following criteria for tendering contracts. Civil works that did not require a high level of technology were channelled to domestic firms. More complex projects were assigned to private and government-owned companies that had accumulated a higher level of expertise because of joint ventures and their active involvement in international projects during the past three decades. The most technically challenging projects are given to international consortia in the form of either joint ventures or technical collaboration agreements with domestic companies. This paper discusses the goals of DORTS during the past several years as they relate to promoting (1) increased domestic input, (2) technology transfer, and (3) a higher level of achievement.
Notes: A case study example of contractor selection methods employed on a large construction project.

London, K and Kenley, R (2000) The development of a neo-industrial organization methodology for describing and comparing construction supply chains. *In*: *Eighth Annual Conference of the International Group for Lean Construction IGLC-8*, 17-19 July 2000, Brighton, UK. University of Sussex.
Abstract: The model draws from theories of industrial economics and supply chain literature, and is an attempt to advance the construction supply chain field through the development of an industrial organizational methodology to describe construction supply chains. Empirical studies have examined the industrial organization of other industries, typically forming descriptions based upon vertical integration and horizontal market concentration. A review of the trends in the supply chain literature indicates there is a need to develop a model to describe the industrial organization of the construction industry through supply chain structure. The merging of the supply chain concept with the industrial organization model as a methodology for understanding the structural characteristics is an important contribution to construction economic theory. The paper begins to develop a language for describing the structure and behaviour of supply chains specific to the construction industry and in so doing suggests a neo-industrial organization approach.
Notes: An application of supply chain theory from outside the construction industry to projects within it.

—— **(2001)** An industrial organization economic supply chain approach for the construction industry: a review. *Construction Management and Economics*, **19**(8), 777-88.

Abstract: Understanding industries in terms of the concepts of chains, clusters and networks is becoming increasingly important in economies around the world. Supply chain management for an individual organization is an emerging field of research in the construction management discipline, but less attention has been devoted to investigating the nature of the construction supply chains and their industrial organizational economic environment. This selected review of construction and mainstream management supply chain literature is organized around four themes: distribution, production, strategic procurement management and industrial organization economics, and highlights the need to develop an industrial organization economic supply chain framework for construction. The merging of the supply chain concept with the industrial organization model as a methodology for understanding firm conduct and industry structure and performance is an important contribution to both construction supply chain and construction economic theory. Much of the industrial organization supply chain literature has tended to focus upon manufacturing industries, where typically firms are permanent organizations. This raises issues as to the differences between industries founded upon temporary compared with permanent organizations. There is potential for the development of an industrial organization methodology applicable to the project based industry. Ultimately industrial organization research seeks to have direct implications for industry performance and government policies.

Notes: In the field of supply chains, comparisons are made between industries founded upon permanent organizations, and those, such as construction, that are temporary.

London, K, Kenley, R and Agapiou, A (1998) Theoretical supply chain network modelling in the building industry. *In:* Hughes, W P (Ed), *14th Annual ARCOM Conference*, September 9-11, University of Reading, UK. Association of Researchers in Construction and Management, Vol. 2, 369-79.

Abstract: Various supply chain management models have emerged in the last two decades in the manufacturing industry. Current supply chain techniques in the construction industry have focused on project-based models, particularly the logistics of materials management. It has been established in manufacturing that the greatest successes have been in those supply chains where strategic alliances have been developed between key parties and longer-term relationships have been developed across the entire network of suppliers at each stage of production. The link between the customers' internal management systems and the external suppliers' management systems has been identified as important to creating successful alliances. Unravelling this complex web of successful buyer-supplier inter-enterprise relationships and dependencies in manufacturing has relied upon the two key causal factors of supplier co-ordination and supplier development. Theoretical SCN modelling in construction needs to consider that the underlying structure of the majority of parties involved within construction projects is largely a network of small to medium enterprises. A theoretical supply chain procurement model is developed that reassesses the fundamentals of organizational structure and explores the potential for future flexible inter-organizational arrangements. The modelling draws upon network sourcing concepts and aims to foster risk sharing and engender stability in contractual chains. The methodology for a future research project is described that will explore supply chains in a small remote construction market.

Notes: Presentation of a theoretical supply chain procurement model that is appropriate for the flexible inter-organizational arrangements that characterize construction projects.

Love, P E D, Skitmore, M and Earl, G (1998) Selecting a suitable procurement method for a building project. *Construction Management and Economics*, **16**(2), 221-33.
Abstract: Building procurement has become a fashionable term with industry practitioners and researchers. It determines the overall framework and structure of responsibilities and authorities for participants within the building process. It is a key factor contributing to overall client satisfaction and project success. The selection of the most suitable procurement method consequently is critical for both clients and project participants, and is becoming an important and contemporary issue within the building industry. The problem, nevertheless, lies in the fact that there has been limited empirical research in this field of study. Postal questionnaire surveys of 41 clients and 35 consultants were carried out, and were used to obtain experience of and attitudes to a variety of procurement methods and the criteria used for selection. The findings indicate that a simple set of the criteria generally is adequate and sufficient for procurement path selection, and that there is a reasonable consensus on the appropriate weighting for each path. Moreover, it is shown that, contrary to expectations, similar clients generally do not have similar procurement needs.
Notes: An approach to the appropriate selection of procurement method.

Lowe, D, Emsley, M and Duff, R (2002) The cost of different procurement systems: a decision support model. *In:* Khosrowshahi, F (Ed), *3rd International Conference on Decision Making in Urban and Civil Engineering (DMinUCE)*, 6-8 Nov 2002, London.
Abstract: Existing research which has attempted to determine differences between the costs of the different procurement routes has consistently aimed to determine a single figure for the difference for projects as a whole. No attempt has yet been made to provide a difference which is project specific. Furthermore, no previous research has determined the cost to the client using any objective method. The absence of such a technique is significant. It means that the client's advisors have no means of providing an objective measure of the cost of following different procurement routes. The client must depend upon the judgement of the advisors, which is based on their own perception of both the project and the different procurement routes, and is hence subject to their opinions and prejudices. This paper reports on the development of a neural network model (ProCost) which is able to determine the total cost to the client of a project, which functions as a decision support tool by enabling the project specific comparison of alternative procurement routes and other strategic decisions.
Notes: As with previous work by these authors, this is of some relevance to the current project, though its concern is to produce a prototype model that predicts the effect of procurement method on *total* project cost, without distinguishing the transactional costs of the procurement process itself.

Marsh, L and Flanagan, R (2000) Measuring the costs and benefits of information technology in construction. *Engineering, Construction and Architectural Management*, **7**(4), 423–35.
Abstract: Information technology (IT) has been widely applied across many economic sectors in order to increase competitiveness and reduce costs. This paper identifies that uptake of IT within construction is low. It is argued that significant barriers preventing construction organizations from investing in IT include uncertainty concerning the identification and measurement of benefits associated with applications. In particular, it is argued that difficulties in quantifying benefits associated with improved information availability and decision making prevent effective IT cost/benefit analysis. Existing approaches to evaluating IT within construction are reviewed. A framework is presented

which identifies metrics by which IT impacts both management and operational processes within construction in order to deliver value. An evaluation methodology tailored to one specific IT application, high-density bar coding in maintenance management, is presented to illustrate the quantification of both the costs and benefits of applying IT.
Notes: The availability of technology, particularly information technology, has a crucial effect on the overall cost of transactions.

Masden, S E, Meehan, J W and Snyder, E A (1991) The costs of organization. *Journal of Law, Economics and Organization*, **7**(1), 1-25.
Abstract: The tenet to which all transaction-cost economists subscribe is that the choice among alternative organizational arrangements turns on a comparison of the costs of transacting under each. To impart empirical content to this fundamental insight, theorists began to relate the incidence of transaction costs to observable attributes of transactions. Although the empirical research to date has been generally supportive of the central transaction-cost propositions, recognition that variations in internal organization costs may also play a role in the decision to integrate exposes an inherent weakness in the nature of these tests. Because of difficulties in observing and measuring transaction costs, analysts have had to rely on estimations of reduced-form relationships between observed characteristics and organizational forms. The results of the research described here indicate that overall organization costs represent about 14% of total costs for the components and activities in our sample. Moreover, these costs were found to vary systematically with the nature of the transaction and that savings from choosing organizational arrangements selectively can be substantial. The broader implications of the study lie in findings regarding the relative contribution of variations in internal and market organization costs to the integration decision. Specifically, we find that as the costs of dealing across a market interface rise, and hence the incentive to integrate, so does the potential for hold-ups in a given transaction, as recent theorists have argued. In addition, however, the evidence indicates that variations in the level of internal organization costs also play an important role in integration decisions. Indeed, the importance of internal organization costs in our results leads us to reassess some of the earlier empirical literature on the determinants of vertical integration.
Notes: This empirical study relies on selecting a limited number of variables and asking the respondents to give an ordinal score to the importance of each factor, related to 74 observations from one firm involved with a shipbuilding contract. These qualitative evaluations are then analysed using econometric methods to test various hypotheses about the integration decision. There is an interesting passage describing the construction industry, which forms a context for the description of shipbuilding. The limitations of this work, most of which are identified by the authors, are connected with using proxies for data instead of real cost data, and with studying only a small sample of decisions from one firm. The idiosyncrasies of the chosen industry are important, such as the complexity of the process and scheduling issues as well as the application of government regulations to defence contracts. There are too many approximations in their data for their conclusions to be reliable, even within the limited parameters of their study. They identify the difficulty of obtaining data as the key obstacle to testing transaction-cost theory.

Matthews, J, Tyler, A and Thorpe, A (1996) Pre-construction project partnering: developing the process. *Engineering, Construction and Architectural Management*, **3**(1/2), 117-31.

Abstract: The use of sub-contracting within the modern construction industry has become commonplace with many main contractors only undertaking the management and co-ordination activities. The reliance on sub-contractors has put much stress on the sub-contractor/main contractor relationship. As main contractors have realized that the greatest potential for cost saving lies with sub-contractors, the prevalence of unfair contract conditions, dutch auctioning and other onerous practices has increased. This paper describes a procurement approach, utilizing limited competition, developed by a top UK main contractor (MC) in order to improve its relationships with sub-contractors. The approach, termed semi-project partnering, was implemented on a commercial development. The approach was supported by research which identified: what MC's employees want from sub-contractors; what sub-contractors want from main contractors; and a study to benchmark MC's performance with that of other main contractors. It was concluded that this approach offers a number of benefits for the client, main contractor, partnering sub-contractors and professional consultants. These included an improved team approach; an improved understanding of the project; more compliant sub-contractor bids; better/closer relationships; more reliable programming; less confrontation; and lower tendering costs. It was also identified through debriefing sub-contractors that sub-contractors were quoting 10% lower than normal due to this approach.
Notes: Partnering is introduced and defined. The research consists of one case study concerning the practices of one main contractor. The conclusions do not flow from the paper as it is only here that Table 1 is introduced. This Table shows that, in one case study, the client, the main contractor and the sub-contractors, but not consultants, felt that partnering will lead to lower tendering costs.

McCaffer, R, McCaffery, M J and Thorpe, A (1983) The disparity between construction cost and tender-price movements. *Construction Papers*, **2** (2), 17-27.
Abstract: This paper describes an examination of the disparity, usually attributed to market conditions, between movements in tender prices and construction costs. Price changes due to market conditions, i.e. price changes independent of underlying cost changes, were represented by the ratio tender-price index to construction cost index. Movements of this ratio were correlated with market indicators such as the level of output or orders, contract value, location and construction type. The results of these studies showed that price changes were highly correlated with changes in output two to four quarters earlier. The degree of correlation and the time lag between movements in output and price changes varied between market sectors. Also, up to 63% of the variability in the market induced price changes of individual contracts was explained by the level of output two to four quarters before tender date, contract value, construction type and location. The feasibility of using regression models, derived from the correlation exercises, to predict tender price levels was tested by determining the model's ability to predict already published tender indices. The test forecasts showed a substantial improvement in comparison to the published forecasts of the same indices. The suggestion is that the use of such models may well have predicted the otherwise unexpected falling trend in tender prices during 1981 and 1982.
Notes: Analytical treatment of tender price movements, offering a means of predicting changes in general pricing levels in the market.

McGill, R E N (1991) Alternative methods of procurement for works of water and environmental-management. *Journal of the Institution of Water and Environmental Management*, **5**(2), 206-11.

Abstract: The paper considers methods of procurement for the construction of water- and sewage-treatment works other than the traditional use of the ICE 5th Edition Conditions of Contract and Bills of Quantities. This is as a consequence of the speed at which schemes are currently required within the water industry. The paper particularly focuses on the use of target contracts, using the Institution of Chemical Engineers' Green Book form, and considers various types of target. It also discusses the form which tender documents may take, and how targets may be calculated. Tendering procedures, including the use of two-stage tenders and negotiated targets, are considered, as are the procedures for cost monitoring and the financial management of the contract. Finally, the role of the engineer and quantity surveyor is considered.
Notes: Consideration of methods of procurement for the construction of water and sewage treatment works.

McKellar, T and Akintoye, A (1997) Electronic data interchange in the UK construction industry. *RICS Research Papers*, **2**(4), 1-30.
Abstract: Effective information exchange is as much a pre-requisite for a successful conclusion to the construction process as it is for any other business function, and perhaps more so. The broad diversity of parties involved, allied to the varied information and documentation requirements throughout the construction cycle from inception to handover ensure its pivotal role. Electronic data interchange (EDI) involves the computer to computer transfer of structured data between trading partners. As a technology, it is reliant on an amalgam of the computer and communication industries, both of which could be said to be at the forefront of modern innovation. EDI facilitates the effective interchange of standard documents, such as invoices and purchase orders, and has been enthusiastically adopted in other industries, notably the automotive, electronics and retail sectors. Construction project documentation in the form of contracts, tender enquiries and bills of quantities are examples of documents passed between parties which may lend themselves to electronic transmission. The objectives of the research were to; examine the approach adopted by other industries towards EDI, establish the extent to which EDI is utilized in the construction industry, identify barriers and benefits of EDI in the construction industry.
Notes: The availability of EDI facilitates the effective interchange of information and thus has an effect on the overall cost of project transactions.

McMaster, R (1992) *Competitive tendering and service quality: a brief examination utilizing an amended transaction cost framework*. Vol. 92-13, Discussion paper. Aberdeen: University of Aberdeen, Department of Economics.
Abstract: The transaction cost framework has been increasingly advocated as providing a more robust analysis of the economic ramifications of competitive tendering than the mainstream neoclassical paradigm. However, this paper contends that the conventional transaction cost framework is analytically flawed and demonstrates a penchant for single exit determinism. Amendments to this framework are posited. This paper postulates, via the amended framework, that competitive tendering in health and local authorities induces an adjustment to existing institutional arrangements and changes the nature of any given exchange. By increasing the role of the price mechanism relational exchange a potential manifestation of the change in the nature of an exchange is an adjustment in the standard of services delivered.
Notes: A criticism, theoretically based, on the use of a transaction cost framework for analysing competitive tendering.

Mee, M S R (2002) *Total cost of procurement*. [Available online from http://www.jppsg.ac.uk/guidances/procost.html]
Abstract: This is one of a series of Procurement Guidance Notes commissioned by the Joint Procurement Policy and Strategy Group (JPPSG). Their aim is to promote the best procurement practices in higher education institutions. While their use is not mandatory, JPPSG expect the highest standards of procurement and accordingly recommend that each institution sets up a formal procedure for the consideration of the Guidance Notes as standard procurement practice, and for their dissemination throughout the institution.
Notes: A guidance note on procurement. The interesting aspect, inherent in the title, is the acceptance that it is the *total* cost of procurement that should be considered. By implication, a particular procurement strategy may be less costly in one phase of the process but more costly in another.

Merna, A and Smith, N J (1999) Privately financed infrastructure in the 21st century. *Proceedings of the Institution of Civil Engineers-Civil Engineering*, **132**(4), 166-73.
Abstract: Growing difficulties and inefficiencies of publicly funded infrastructure have led to a return of pre-1900s methods of private financing. Though investors are naturally wary with no recourse to public funds, increasing sophistication of financial engineering techniques means it is now possible to match finance requirements with projected cash flows. This paper reviews the state of project finance and outlines the instruments and techniques of financial engineering. It also looks at the risks associated with privately financed infrastructure projects in the future.
Notes: Descriptive account of the problems associated with funding PFI and PPP projects. No assessment of the costs tendering is included.

Micelli, E (2000) Mobilizing the skills of specialist firms to reduce costs and enhance performance in the European construction industry: two case studies. *Construction Management and Economics*, **18**(6), 651-56.
Abstract: Analyses are made of procurement strategies capable of leading to cost reduction and higher quality through the contribution of specialized firms. An exploration is made of the literature concerning the procurement of manufactured goods and progress by analysing two case studies: the East Bridge of the Storebælt link and the Grand Canal Maritime Bridge. Both of these projects adopted procurement systems that led to a learning process capable of enhancing performance and reducing overall costs. It is evident that the success of learning-oriented procurement strategies relies on two major conditions. First, the client's tender must be issued with an open design demanding an active contribution from the firms involved in the construction process. Second, the firm's bidding for the job must be able to manage two distinct sets of skills: the technical skills relating to a specific production process and the ability to connect these skills to the client's needs through a co-design process, or what is known as the strategic intermediation function.
Notes: Examines procurement systems and investigates two case studies where learning-oriented procurement strategies were adopted. These lead to a reduction in production costs. No real consideration of the costs of procurement.

Miller, C J M, Packham, G A and Williams, T (1999) Re-defining sub-contracting: reducing transaction costs? *In:* Hughes, W P (Ed), *15th Annual ARCOM Conference*, September 15-17, Liverpool John Moores University, UK. Association of Researchers in Construction and Management, Vol. 2, 655-64.

Abstract: It is maintained that partnering can arguably reduce ex-post transaction costs within the construction process. These costs primarily arise due to a lack of harmonization between contracting parties. Historically, this relationship has been transactional in nature, with both parties seeking to secure value added at minimal cost. Despite this fact, evidence suggests that mutual co-operation that can supersede a traditional cost led approach offers new hope for prosperity in the construction industry. This paper offers preliminary research highlighting that the competent implementation of strategic partnerships based upon trust can only reduce transaction costs if the small sub-contracting firm is fully integrated into the process. The paper concludes that traditional approaches and new practices will, if they continue to facilitate contractor opportunism, encourage small sub-contracting firms to seek alternative markets instead of enabling mutual co-operation to reduce the transaction costs of all stakeholders involved in the construction process.
Notes: The costs considered in this paper are *ex-post* transactions costs, after tendering. The hypothesis is that partnering, if properly implemented (and in this instance this means the integration of sub-contractors into the arrangement), can reduce the potential for costs in these areas by attenuating the need for opportunistic behaviour.

Mirza and Nacey Research (2001) *Consultants' Performance.* Arundel, West Sussex: Mirza and Nacey Research.
Abstract: This report examines how much consulting firms earn in fees; what the total market is worth; and how these benchmarks have changed. Consultants' Performance builds on the 1999 report Architects' Performance, and has developed to include comparative data for three different professional consultant groups. The research reported here is based on the long-established Architects Quarterly Workload Survey together with two newer regular surveys amongst quantity surveyors and consulting civil engineers specially undertaken by Mirza and Nacey. This research is aimed at being of immediate practical use to architectural, quantity surveying and civil engineering consultants, in both the private and public sectors.
Notes: Some useful information of fee levels and fees earned by consultants, from which can be extrapolated a figure for total fee income.

Moore, D R and Dainty, A R J (2001) Intra-team boundaries as inhibitors of performance improvement in UK design and build projects: a call for change. *Construction Management and Economics*, **19**(6), 559-62.
Abstract: The success of the design and build (D&B) procurement route could be undermined by issues arising from the rigid professional cultures of individual participants within project workgroups. These have the potential to inhibit the achievement of a key espoused benefit of D&B procurement, i.e. that it promotes the integration of the design and construction processes for improved project performance. Cultural non-interoperability is identified as a significant potential barrier to effective change management, and to the achievement of innovation within the design and construction processes. This note argues that project responsibilities, which currently are delineated along professional identity lines, produce design and construction solutions that fail to fulfil the potential of D&B procurement. It is suggested that addressing cultural interoperability will require a fundamental and long term reshaping of the industry's structure, beginning with the professional bodies and the higher education system that underpins their future membership.
Notes: Theoretical paper highlighting the socio-cultural barriers to integrated working.

Murray, M D and Langford, D A (1998) Construction procurement systems: a linkage with project organizational models. *In:* Hughes, W P (Ed), *14th Annual ARCOM Conference*, September 9-11, University of Reading, UK. Association of Researchers in Construction and Management, Vol. 2, 544-52.
Abstract: This paper constitutes a literature review undertaken at the start of a two and a half year EPSRC-funded research project. As such, its purpose is to present the details of the research concerning construction procurement and project organizational design. The paper shows that the post-Latham construction industry provides several new developments (client power, partnering, concurrent engineering etc) which are altering the construction project process, and therefore prove worthy vehicles for investigation into project organizational structures.
Notes: Discussion of post-Latham procurement developments.

Nahapiet, H and Nahapiet, J (1985) A comparison of contractual arrangements for building projects. *Construction Management and Economics*, **3**(1), 217-31.
Abstract: This paper considers contracts from an organizational perspective, comparing the major forms of contracts available for building projects and examining the factors influencing their selection. The analysis is based on the findings of a study of ten building projects, six in the USA and four in the UK, together with the results of a survey of those prominent in the industry. A comparison of five different contractual arrangements indicate that they establish different patterns of responsibilities and relationships between clients and the various parties involved in building projects. In so doing, they are regarded as offering clients differing combinations of expertise, risk, flexibility and cost. For the projects studied, three factors were found to be related to contract selection: the characteristics of clients, particularly their experience and expertise in construction; the level of performance required by clients and the construction complexity of projects. These findings, together with previous research, suggest that it is unlikely that there is one "best" form of contract for building projects. Rather, which is the appropriate contractual arrangement varies according to the particular set of project circumstances, especially the type of client, his time and cost requirement and the characteristics of the projects.
Notes: A case-study based examination of the appropriateness of five different procurement methods.

Naim, M M, Childerhouse, P, Disney, S M and Towill, D R (2002) A supply chain diagnostic methodology: determining the vector of change. *Computers and Industrial Engineering: International Journal*, **43**(1/2), 135-57.
Abstract: The paper presents a guide to conduct a supply chain oriented business diagnostic or "health check" termed Quick Scan. The Quick Scan is a systematic approach to the collection and synthesis of qualitative and quantitative data from a supply chain. The Quick Scan approach is the initial step in a generic methodology for identifying the change management opportunities in the supply chain. The paper highlights the need for adequate analysis of the supply chain before embarking on the route of information and communication technology implementation. As well as being of operational benefit to specific companies the Quick Scan may also be utilized to develop generic research models of supply chain change. The results of the application of the methodology to twenty European automotive supply chain value streams are evaluated. The results show that only 10% of automotive supply chains are close to the goal of an integrated supply chain, whilst only 30% of supply chains exhibit much good practice. The remainder are still struggling to

implement `lean" production techniques, which is a common pre-occupation with many automotive industry executives. The Quick Scan is thus able to advise companies in terms of the direction and magnitude of change required in their supply chains.
Notes: An introduction to a diagnostic tool for evaluating the effectiveness (in terms of integration) and efficiency (in terms of "leanness") of supply chains. The sample was twenty European automotive supply chains.

Naoum, S G (1994) Critical analysis of time and cost of management and traditional contracts. *Journal of Construction Engineering and Management, ASCE,* **120**(4), 687-705.
Abstract: This paper presents some findings of a PhD research program conducted by Naoum in 1989, which sought to investigate whether the means of procurement influenced project performance. Ten factors were identified to measure project performance: (1) Pre-construction time; (2) construction time; (3) total time; (4) speed of construction; (5) unit of building; (6) time, overrun; (7) cost overrun; and client satisfaction with (8) time; (9) cost; and (10) quality. A theoretical framework was used to assist in comparing project performance in a case study sample of 39 management contracts and 30 traditional contracts. From the evidence, it must be concluded that the solution to the problems facing the construction industry lie in neither the management nor the traditional system. To achieve project success, the parties need to match the various organizational forms to the client's characteristics, criteria, and priorities with respect to time, cost and quality. The statistical analysis suggests that management contracting performs significantly better in some respects than traditional contracting, in particular, when time was important in the contract and when the project was highly complex. However, there is not enough evidence to support the view that management contracting can reduce the overall building cost, or that the system can increase the standard of quality.
Notes: A comparison between "traditional" and "management-based" procurement methods. No consideration of transaction costs. Ten indicators of project performance were measured, but there was no consideration of the actual costs of procurement or subsequent transaction costs.

National Audit Office (1997) *Bridgend and Fazakerley Prisons: National Audit Office Report*, London: National Audit Office.
Abstract: (From the Press Release) The Private Finance Initiative contracts for new prisons at Bridgend and Fazakerley are expected to provide the Prison Service with 1400 new prisoner places. The contracts, which require the contractors to build the prisons and to maintain and operate them for 25 years, are expected to cost £513 million – an estimated aggregate saving of 10% when compared with building prisons using public finance and contracting out the operating to the private sector. A key feature of these PFI contracts is that the private sector contractors – Securicor and Costain at Bridgend and Group 4 and Tarmac at Fazakerley – will be responsible during the 25-year contract periods for all custodial services and ancillary services such as catering, prisoner education and onsite medical facilities. Compared to traditional procurement methods, the PFI solutions will provide a faster construction period, innovative forms of design and operational methods and transfer major construction, maintenance and operational risks to the private sector. These were the first PFI contracts involving the construction of buildings. The Prison Service identified lessons that they have applied in subsequent prison procurement. They have also sought to establish procedures for resolving quickly issues with their contractors and opportunities for developing the projects in a mutually beneficial way. Sir John said that

this partnering approach is important, given that the contracts run for 25 years. The National Audit Office found that: the Prison Service made a strategic decision to let the Bridgend contract to Securicor and Costain and the Fazakerley contract to Group 4 and Tarmac because they saw operational and constructional risks in letting two major PFI contracts to any one bidder, including that Securicor, although experienced in providing security services, did not have previous experience of managing a prison and that Costain were experiencing financial difficulties. The Service also hoped to encourage the developing market in prison procurement under the PFI, leading to improvements in value for money on future PFI prison contracts; the lowest bidder, the Securicor/Costain consortium, told the Prison Service when they submitted their bid that additional savings could be delivered through economies of scale in the event of their being awarded both contracts. The Service did not ask them, or other bidders, to quantify such possible savings as anticipated when the Fazakerley contract was awarded the Service have subsequently contracted with Tarmac for the construction of seven further house-blocks at various prisons using a new offsite technique which Tarmac developed for the Fazakerley project. This technique enables long term secure accommodation to be delivered between 6 and 11 months faster than traditional methods; the extent to which value for money will be secured during the 25 year contract periods will depend on how the contractors' prices compare with charges by other providers for similar services. There is no requirement, however, for the Bridgend and Fazakerley contract prices to remain in line with the charges of other suppliers during the contract periods; and the cost of advisers and consultants of £1.6 million reflected the work required for a new and complex method of procurement but was more than double the estimate of £0.6 million because the Prison Service had no previous experience of a major PFI contract, underestimated the amount of work that was necessary, and did not expect to have to re-tender during the procurement. The Prison Service, however, believes that their expenditure on legal advice in these first PFI projects will have long term benefits to their subsequent PFI projects. The contract for their financial adviser, Lazards, was let without competition. The Report recommends that departments should: (1) consider, where more than one contract is being let simultaneously, whether bidders would be prepared to reduce their proposed contract prices for the benefit of working on more than one contract. This would need to be balanced against other benefits which individual bidders could provide and the possibility that Departments might be exposed to a greater degree of risk where there is only one provider; (2) consider including in very long term contracts a facility for contract prices to be compared at periodic intervals with those charged by both public and private sector providers of comparable services and for contract prices to be adjusted if they are more expensive than the charges of other providers. Departments will need to take account of any upwards price revisions which contractors may require in return for agreeing to this facility; (3) seek to ensure, before signing a PFI contract, that the contract unambiguously reflects their understanding of how risks are to be allocated between the signatories; (4) normally appoint advisers for PFI projects after competition, even where an adviser has provided preliminary advice on PFI matters; (5) gather information which will enable them to establish an accurate budget for each adviser's costs at the earliest practicable stage in a PFI procurement process. Particular care should be taken with the preparation of budgets for legal work as this is likely to be one of the major items of expenditure in connection with the letting of a PFI contract; and (6) seek to develop a cooperative relationship with the supplier, in order to resolve issues quickly and deliver benefits such as cost reductions, improved quality and innovative solutions.

Notes: This is an interesting report on the process for letting two PFI projects. The following are of particular relevance. (i) On page 55, is a table breaking down the client's costs of procurement. The cost of procurement was £1.55 million, compared with a project price of £636m. These costs were incurred in paying advisors. (ii) The time to let the contracts was longer than expected (25 months compared to an expected 16 months). This raises the question of how to cost the time required for various forms of procurement. (iii) There were initially six bidders but only five survived to the final stages. However, ten provided pre-qualification data (it must be remembered that there were two projects). (iv) The report says nothing about the costs to the bidders. (v) A problem with PFI will be that there is often no indication of the value of the building work.

Nettleton, B (2000) Best value and direct services. *Proceedings of the Institution of Civil Engineers-Municipal Engineer*, **139**(2), 83-90.
Abstract: The effects of compulsory competitive tendering (CCT) on local authorities: organizational and managerial change have forced competitiveness on price and productivity; Direct Labour Organizations (DLO) have improved performance and held their own; but CCT has also brought fragmentation of services and wastefulness of resources. Direct services provision through DSOs, if updated and restructured, is presented as a good option for both best value and integration of services.
Notes: Charts the changes to Local Authority procurement following the Local Government Act 1999 and its replacement of mandatory CCT (compulsory competitive tendering) with Best Value. Explains Best Value. Mentions the negative effects of CCT as "fragmentation" and "duplication" in terms of monitoring, supervising and inspecting" (p.88).

Ng, T S, Kumaraswamy, M M and Chow, L K (2001) Selecting consultants through combined technical and fee assessment: a Hong Kong study. *In:* Akintoye, A (Ed), *17th Annual ARCOM Conference*, 5-7 September 2001, University of Salford. Association of Researchers in Construction and Management, Vol. 1, 639-48.
Abstract: Many clients increasingly employ a competitive bidding approach for recruiting construction consultants. However, the concept of relying on the bid price alone is problematic as a consultant submitting the lowest bid may not necessarily be able to complete the work satisfactorily, and any errors in design or supervision may in turn cost the client many times the savings accrued from a low consultant fee. A proper consultant selection process, which takes into account other quality-based criteria, is therefore necessary to ensure the quality of the consultants appointed. This paper examines a Combined Technical and Fee Assessment (CTFA) approach being used in Hong Kong (HK), and discusses the weaknesses of the current CTFA approach. The initial results indicate that the disparity in the usage and relative importance of assessment criteria between various clients and the over-reliance on expert judgement in assigning the weightings are the major concerns of the current CTFA approach.
Notes: A description of a method by which consultants' tenders may be assessed using criteria that incorporate quality as well as price.

Ng, T S, Skitmore, M R and Smith, N J (1999) Decision-makers' perceptions in the formulation of pre-qualification criteria. *Engineering, Construction and Architectural Management*, **6**(2), 155-65.
Abstract: Contractor pre-qualification involves the establishment of a standard for measuring and assessing the capabilities of potential tenderers. The required standard is

based on a set of pre-qualification criteria (PQC) that is intended to reflect the objectives of the client and the requirements of the project. However, many pre-qualifiers compile a set of PQC according to their own idiosyncratic perceptions of the importance of individual PQC. As a result, sets of PQC, and hence pre-qualification standards, vary between pre-qualifiers. This paper reports on an investigation of the nature of the divergences of the perceived importance of individual PQC by different groups of pre-qualifiers via a large-scale empirical survey conducted in the UK. The results support the conclusion that there are significant systematic differences between groups of pre-qualifiers, with the individual PQC that contribute most to the differences being the method of procurement, size of project, standard of quality, financial stability, project complexity, claim and contractual dispute and length of time in business.

Notes: This paper deals with the criteria considered important by clients and consultants for the assessment of contractors' abilities to do a job, i.e. pre-qualification criteria.

Nkado, R N (1995) Construction time-influencing factors: the contractor's perspective. *Construction Management and Economics*, **13**(1), 81-9.

Abstract: The results of a preliminary survey of factors affecting construction time is described. The objective of the survey which was conducted in the UK was to prioritize factors which are taken into consideration by accomplished contractors in planning the construction time of buildings. A significant degree of consistency in the ranking "time-influencing factors" was found. The most important factors are apparently those which can readily be identified or deduced from the project information and whose impact on construction can generally be assessed explicitly by mathematical and judgement analyses.

Notes: The paper is based on a preliminary survey of factors affecting construction time.

Nunn, D (1999) What has the Egan initiative achieved? Benchmarking Egan. *Contract Journal*, **398**(6227), May 26, 14-19.

Abstract: Looks at progress of the Egan initiative six months on, based on telephone interviews with 100 of the industry leaders present at the launch. Charts and comment on the industry's response: actions since Egan; benchmarking performance; provision for the workforce; women and ethnic minorities; and on the government's clients and partnering, tendering and contracts.

Notes: A post-Egan Report survey of "industry leaders" which gives their opinions on the changes six months after its publication.

Odusote, O O and Fellows, R F (1992) An examination of the importance of resource considerations when contractors make project selection decisions. *Construction Management and Economics*, **10**(2), 137-51.

Abstract: This research investigates the factors which building contractors consider when making project selection decisions and the processes they follow when making such decisions. The research is based on the premise that resource consideration is the most important factor for contractors when making project selection decisions. Literature was used to identify those factors thought to be most influential. The extent of the diversity of opinion disclosed indicated the necessity to obtain further information which was gathered from contractors by use of postal questionnaires and interviews. The information was analysed by ranking techniques and the results compared by Spearman rank correlation coefficient. The results obtained indicated the strengths of the factors identified. Spearman rank correlation coefficient showed a positive correlation for the results obtained from the findings of the literature study and those obtained from the survey. A non-weighted model of contractors' tendering decision process was developed. The model shows the inter-relationship that exists between client and contractor decisions. The model also illustrates the effect of such decisions on the actions of both clients (their representatives) and contractors. The results obtained from the survey shows that, "the ability of the client to pay for the cost of work" is the most important factor contractors consider when making project selection decisions.

Notes: Development of a model of contractors' tendering decision process derived from survey research.

Pasquire, C L (1994) Early incorporation of specialist M&E design capability. *In:* Rowlinson, S (Ed), *CIB W92 Procurement Systems Symposium*, 4-7 December 1994, Hong Kong. The Department of Surveying, Hong Kong University, Vol. 1, 259-68.

Abstract: The M&E installation is systems oriented and therefore places considerable design demand upon the installer. A well planned design will involve the M&E specialist at an early stage to overcome the problems relating to co-ordination and integration of the M&E systems with each other and with the building structure and fabric. This paper highlights some of the problems experienced when the M&E design is undertaken late and /or the implications of its incorporation are not appreciated by the design leader. The paper follows this discussion with examples of some practices that were successfully overcome and suggests the consideration of some strategic factors within project procurement would simplify and rationalize the design. The ultimate aim of changing the procurement approach is to make substantial cost savings.

Notes: Encourages the early incorporation of the M&E design into the project design. As much of this is done by the installer, the upshot is the need for a procurement system that integrates the M&E specialist contractor as early as possible.

Pasquire, C L and Collins, S (1996) The effect of competitive tendering on value in construction. *RICS Research Papers*, **2**(5), 1-32.

Abstract: This paper reports on an examination of the performance of single-stage competitive tendering in terms of delivering value. The research was undertaken as it has been indicated that value for money may not be the only outcome of competitive tendering. The paper examines the three areas of: pre-qualification, existing tendering practice, negotiation. The research introduced issues such as the use of cover prices and negative margins, and surveyed both contracting companies and client organizations. The analysis revealed surprising attitudes towards value and competition, particularly among client bodies. The work has particular relevance, as it was undertaken just prior to and immediately after the publication of the Latham Report (1994), and investigated some of the issues which the Latham Report considered. The main conclusions of the work are that some elements of competition do indeed encourage value for money, but that there are many procedures within the competitive tendering process that adversely affected value. Both types of elements are identified. Although ostensibly confined to the UK, the work has implications within any national industry that adopts similar competitive tendering procedures.

Notes: This paper is about the effect of competitive tendering on value in construction. In particular, it looks at: (i) The maximum length of tender list for traditional contracts (pp8-9). (ii) The maximum length of tender list for D&B contracts (surprisingly many) (p9). (iii) A little on the cost of pre-qualification (p7) and the method of recovery (pp7-8). (iv) A little on cost of tendering for D&B (quoting Latham) (p9). (v) Some information on cover pricing showing that 65% of the contractors surveyed would be prepared to submit a non-competitive price (p10). It does not have information on the incidence of cover prices. A problem with the data is that the survey on which it is based contained only 26 contractors (out of a possible 50) and 12 clients (out of a possible 25).

Pigg, D R (1993) An introduction to compulsory competitive tendering for professional engineering services. *Proceedings of the Institution of Civil Engineers-Municipal Engineer*, **98**(2), 69-78.
Abstract: Provides background information on the introduction of compulsory competitive tendering to construction related services in general.
Notes: An example of a number of papers and articles that appeared after professional design services, particularly those of in-house public sector design teams, were exposed to compulsory competitive tendering. To some extent these pieces are dated (following the Best Value requirements of the 1999 Local Government Act).

—— **(1995)** Management for survival in a competitive environment. *Proceedings of the Institution of Civil Engineers-Municipal Engineer*, **109**(1), 21-8.
Abstract: Addresses compulsory competitive tendering (CCT) from the perspective of a local authority wishing to retain a significant in-house consultancy division, delivering the required services at the required quality at the right price, within a competitive environment.
Notes: This paper offers advice to local authority engineering departments about how to retain a significant in-house engineering consultancy division in the face of Compulsory Competitive Tendering, which was being introduced at the time for consultancy services, having already been adopted for manual work. The paper is about how such a department can be competitive with private sector consultants. There is no indication about the costs of preparing tenders or any other aspects of the business processes. It would have been interesting to compare in-house management costs with the costs of going to the market.

Poh, P S H and Horner, R M W (1995) Cost-significant modelling: potential for use in South-East Asia. *Engineering, Construction and Architectural Management*, **2**(2), 121-39.
Abstract: A rich variety of cost models is used in the world's construction industries. In countries exposed to British practice, the use of traditional bills of quantity is common. Elsewhere, bills of quantity may not be used at all. This paper briefly reviews the nature and purpose of cost models both in the UK and in South-East Asia. It explains how the principle of cost-significance can lead to a simplified method of measurement, which is both well-structured and a sufficiently accurate half-way house between traditional bills and a single lump sum. By way of example, the derivation of a cost-significant model for student hostels in Singapore is presented. Representing no more than a first step, the problems still to be resolved are outlined. Nevertheless, the techniques seem to hold much promise for the future, and others are encouraged to explore where they might most effectively be applied.
Notes: Cost-significant estimating means that only the most significant items are priced. 80% of the cost lies in 20% of the items. Pricing only 20% of the items gives an estimate that is accurate to 5%. Thus, in theory, the technique reduces estimating effort, but the authors have not actually quantified the estimating effort.

Pokora, J and Hastings, C (1995) Building partnership: team working and alliances in the construction industry. *In:* Harlow, P (Ed), *Construction Papers, No. 54*, Ascot: Chartered Institute of Building.
Abstract: This paper explores the ideas of partnering, team working and collaborative working and their relevance in construction, drawing on research and practical experience from diverse countries and industry, to make the case for the UK construction industry to embrace rapidly partnership conceits if it is to thrive against international competition in the next century.
Notes: Written by two management consultants, extolling the virtues of partnering and strategic alliances, which they can help set up, as consultants. Clearly a piece designed to market their consultancy services. It is based on hearsay and selective quotations from whatever source happens to support their view. They quote from an unspecified Australian Government report which, in turn, quotes some unspecified surveys that show that partnering results in less adversarial relationships, improved resource planning, increased openness, increased trust, improved safety, fewer errors, improved quality, increased contractor profitability (up to 10%), reductions in schedules (around 7%), reductions in engineering costs (up to 10%). Work from Arizona Department of Transport is also quoted, which claims that ADOT pioneered the partnering concept in 1991. Having built 120 projects this way, with a total bid amount of $300m, they claim the average time saved is 12%, the cost saving (over bid amount) is 19% and that added to savings through V.E. exercises, the savings come to 19.5%. The authors are not aware of any systematic work in the UK that estimates the benefits of partnering, but suggested that Haden Young were making good progress in pushing for this kind of relationship and could show good cost and time savings. Overall, this is not a research paper, but a thinly disguised advertisement for a consultancy service.

Quah, L K (1991) Perceptions and management of the risks in tendering for refurbishment work. *Building Research and Information*, **19**(6), 256-359.
Abstract: This paper presents the results of a survey to investigate a large body of UK Contractors' claims that they were subjected to higher risks in refurbishment work; this higher risk exposure being reflected in the greater variability in tender bids for such work.
Notes: As a result of the growing need to conserve existing buildings and to modernize, with emphasis on accommodating modern building services and methods, refurbishment is becoming a major sector in the UK construction industry. Although it is generally perceived as having higher risks, it was acknowledged to be very lucrative as a result of the profits derived purported high risk exposure, poor and unsatisfactory state of tender documents, providing scope for claims and variation in orders, an opportunity for improvement of profits. Generally, project sizes were smaller in value and of a shorter duration, but an extension of time is readily granted for delays, thus in most cases final accounts exceeded the original contract sum. Tender adjudication meetings were attended by groups of 6-9 persons. The inaccuracy of the net estimate and delays to projects were perceived to carry a higher risk. Discounts and the scope for conversion of provisional sums into items of work were considered when deciding the mark-up price. In the management of risks and bidding strategy intuition was commonly used for decision making and risk allowances were usually incorporated in the tender by lump sum additions to net cost, or by higher mark-up. Discounts and scope for improvement of predicted profitability were often used as bidding strategy. The average success rate of tenders was almost 12% and higher for specialist firms. The perceived higher risk in refurbishment work is reflected in the higher risks in

estimating and tendering for such work. However, in practice they may not be as high as perceived. The paper refers to some factors that influence the tender price, such as the proportion of fixed costs within the tender, and the discounts and scope for conversion of provisional sums into items of work when deciding the mark-up price to be applied. It does not emphasize the costs of tendering but rather the risks.

—— **(1992)** Competitive tendering for refurbishment work. *Building Research and Information*, **20**(2), 90-5.
Abstract: The investigation was undertaken as part of a larger study to evaluate a large body of UK contractors' claim of higher risk exposure in refurbishment work. The bidding set or number of bidders in a project; the bid spread or "money left on the table" by the lowest bidder and the success rate and competitive mix were used to gauge the competition in tendering. On balance, it may be concluded that the probability of winning a competitive tender for refurbishment work is higher than in new build work.
Notes: Of interest here is the point that the costs of tendering for refurbishment work are exacerbated where there are large tender lists resulting in considerable "winners curses" in the form of "money left on the table".

Rahman, M M and Kumaraswamy, M M (2002) Joint risk management through transactionally efficient relational contracting. *Construction Management and Economics*, **20**(1), 45-54.
Abstract: The appropriate contracting method and the contract documents for any construction project depend on the nature of the project, but an appropriate contracting method coupled with clear and equitable contract documents do not by themselves ensure project success where people work together in the face of uncertainty and complexity with diverse interests and conflicting agendas. The attitudes of the contracting parties and the co-operative relationships among the project participants are important for successful project delivery. These are examined in the light of transaction cost economics and relational contracting (RC) principles. It is found that RC may well be a useful route towards reduced transaction costs, while also fostering co-operative relationships and better teamwork that in turn facilitate joint risk management (JRM). The usefulness of the latter is reinforced by relevant observations from a recent Hong Kong-based survey, followed by a case study in Mainland China. A basic model is conceptualized for improved project delivery via JRM. This is also seen to be reinforceable by further transactional efficiencies that can be achieved through other RC-based approaches, such as partnering or alliancing.
Notes: Rather speculative and unquantified, this paper suggests that relational contracting is a "useful route towards reduced transaction costs." Presumably this refers to *ex-post* transactions costs. The research base is a single case study in China, reinforced by an earlier attitudinal survey conducted in Hong Kong.

Rankin, J H, Champion, S L and Waugh, L M (1996) Contractor selection: Qualification and bid evaluation. *Canadian Journal of Civil Engineering*, **23**(1), 117-23.
Abstract: Historically the selection of contractors in North America was almost exclusively based on the lowest tendered price. Owners utilized competitive tendering for its simplicity and fairness in the award process. More recently, owners have reconsidered the use of low price as the basis for selecting contractors and therefore have attempted to modify the traditional low bid system with qualification and evaluation clauses. Qualification and evaluation clauses have caused considerable controversy in the public sector, where

contractors are concerned about the legitimacy of tendering procedures when public funds are used. Although contractors agree that the system needs improvement, there are conflicting views on what should be done. This paper provides a summary of traditional tendering procedures and illustrates two possible modifications to the contractor selection process. The intended audience is practicing professionals, particularly those in the public sector where the objectivity of the tendering process is under such scrutiny. Evaluation is selected as the direction with the most potential for improvements. The evaluation option is analysed by examining its legal implications through a discussion of recent case law. Finally, recommendations for implementing an evaluation system are discussed to address the issues identified through the legal discussion and from the opposing views in the construction industry.
Notes: An American example of attempts to modify a traditional low-bid selection system with the introduction of "qualification and evaluation" clauses.

Rowlinson, S and Raftery, J (1997) Price stability and the business cycle: UK construction bidding patterns 1970-1991. *Construction Management and Economics*, **15**(1), 5-18.
Abstract: Problems of competitive pricing and strategic management in the construction industry are discussed. A statistical analysis of tender spread patterns over the period 1970-91 shows that changing market conditions influence levels of risk exposure and in turn affect the establishment of a market-generated "going rate" for construction. A pattern of increasing stability of pricing is identified during the 1980s, and this pattern is linked to developments in the strategic management of contracting organizations. Despite trenchant criticism of the sealed bid as a method of price determination, the industry's price levels do respond relatively quickly to changed economic conditions.
Notes: A study of bid price patterns over a 21-year period in the UK. No assessment of the costs of tendering.

Ray, R S, Hornibrook, J, Skitmore, M R and Zarkada-Fraser, A (1999) Ethics in tendering: a survey of Australian opinion and practice. *Construction Management and Economics*, **17**(2), 139-53.
Abstract: The main issues in the philosophical foundations of ethics and tendering ethics are outlined, and an introduction is provided to the Australian codes of tendering practice. A questionnaire survey is then described which sought to ascertain the extent to which ethical behaviour in tendering is supported and practiced in Australia. The results of the survey indicate that most companies support the use of codes of tendering; defend the right of withdrawal of tenders; disapprove of bid shopping, cover pricing and union involvement in the tendering process, and support the principals' right to know what is included in a tender as well as the self-regulation of the tendering codes. It is also shown that most companies have developed, and follow, idiosyncratic ethical guidelines that are independent of, and often contrary to, the nationally prescribed codes. The conclusions recommend a need for the development of a theoretical frame of reference that can be tested through a more detailed empirical approach to the development of future ethical prescriptions in the field.
Notes: A study of ethics in Australian tendering practice, including a survey with 40 respondents. No mention of the costs of tendering.

Reid, G C (2000) *Rethinking transaction cost economics: a case study of the extension of compulsory competitive tendering to municipal leisure management.* Glasgow: Glasgow Caledonian University Faculty of Business.
Notes: Focuses on compulsory competitive tendering, not on costs of tendering

Ren, Z, Anumba, C J and Ugwu, O O (2001) Construction claims management: towards an agent-based approach. *Engineering, Construction and Architectural Management*, **8**(3), 185-97.
Abstract: Disputes are now considered endemic in the construction industry. They often arise from the poor resolution of claims in the course of construction projects. Efforts have been geared towards reducing the incidence of claims. These efforts are of two kinds: those that seek answers from basic principles and legal issues at the pre-construction phase and those that attempt to solve the problems through claims management procedures at the construction phase. This paper reviews the developments in claims management and highlights the deficiencies in current claims management approaches. It focuses on the need for improvement of the efficiency of claims negotiation and suggests the use of multi-agent systems as an approach to achieve it. The potential benefits of the suggested approach are discussed in the concluding section of the paper.
Notes: The paper introduces a "multi-agent" approach to tackling claims. The cost of claims is an important element of the *ex-post* transactions cost, and therefore the paper is of interest, however it provides no quantitative evidence of the "potential benefits" of the system.

Runeson, G (1988) An analysis of the accuracy of estimating and the distribution of tenders. *Construction Management and Economics*, **6**(4), 357-70.
Abstract: This paper is a statistical analysis of the distribution of tenders and how it can be derived from the distribution of estimators' cost rates. It derives equations to explain the distribution of unit and elemental rates which are found to depend on the contribution of the task to the total cost, labour content and procurement method. It further demonstrates that while there is considerable co-variance between unit rates, there is no such interdependence between either elemental or sub-contractor rates. This means that the central limit theorem can be used to forecast the distribution of tenders.
Notes: The main interest of this author's work is construction price determination, and the efficacy of traditional tendering methods in doing this.

Runeson, G and Raftery, J (1998) Neo-classical micro-economics as an analytical tool for construction price determination. *Journal of Construction Procurement*, **4**(2), 116-31.
Abstract: This paper demonstrates that neo-classical micro-economics is a suitable tool of analysis for the building industry, but that its acceptance requires some change in the way construction economists look at the industry and its output. It examines the history of economics-based discussions of construction price determination by considering the construction economics literature since 1974, and briefly outlines the mainstream economic model of price determination in the context of construction projects. It answers the main arguments which have been advanced against the application of the micro-economic model to construction price determination and demonstrates that this model is not only relevant to construction but also the gains in informative content.
Notes: The author's interest lies in setting construction tendering methods in the context of economic theory.

Runeson, G and Skitmore, R M (1999) Tendering theory revisited. *Construction Management and Economics*, **17**(3), 285-96.
Abstract: The content, origin and development of tendering theory are considered in terms of a theory of price determination. Tendering theory determines prices and is different from game and decision theories, and in the tendering process, with non-cooperative, simultaneous, single sealed bids with individual private valuations, extensive public information, a large number of bidders and a long sequence of tendering occasions, there develops a competitive equilibrium. The development of a competitive equilibrium means that the concept of the tender as the sum of a valuation and a strategy, which is at the core of tendering theory, cannot be supported, and that there are serious empirical, theoretical and methodological inconsistencies in the theory.
Notes: A theoretical treatise on economic theory underlying tendering.

Russo, J G (1981) Construction marketing in the USA: a viable system for success and control. *Construction Papers*, **1**(2), 29-35.
Abstract: While marketing has been an important part of doing business for most industries for centuries, in the construction industry this has not been the case. There have been several major changes affecting construction in recent years, however, that make successful operation of construction companies more difficult. The industry's responses to these changes have been diverse, but the response that has shown the greatest potential impact is the implementation of formal marketing systems. Construction has been slow over the years in adopting good and formal marketing systems. Competitive bidding (tendering), age and size of construction companies, and lack of understanding of marketing techniques have contributed heavily to this failure to apply marketing systems. Understanding the five major areas critical for application of a complete marketing system for construction firms can be the first step toward a more successful and controlled construction operation.
Notes: This is a general paper about change in the construction industry and marketing.

Sewell, G (2001) Peering behind the mask of trust: Between trust and control in contemporary organizations. *University of Melbourne, Department of Management Working Paper in Human Resources Management, Employee Relations and Organizational Studies*, **5**, 1-24.
Abstract: This article explores the relationship between trust and control under the conditions of teamwork found in contemporary organizations. The orthodox approach to organizational trust is characterized as a matter of "economics". Under the rubric of the economics of trust, any mention of control is inimical. This article argues, however, that we should consider relations founded on control, not as the antithesis of relations founded on trust, but as their antimony. In this way control and trust can be seen as being in a state of constant tension with each other, rather than as mutually exclusive social processes. The inspiration for this position comes from a re-evaluation of Michel Foucault's two dimensions of "bio-power" – the "genealogical" and the "epochal". Adopting a position which combines these two dimensions allows us to peer behind the mask of trust and frame our discussion of the relationship between trust and control under conditions of teamwork in such a way that it avoids the normative, instrumental and universal characteristics of the economics of trust. Finally, on the basis of this discussion, the article argues for the development of a micro-ethics of trust where the personal conduct is liberated from the normalizing force of an economics of trust.

Notes: A theoretical discourse about trust – a notoriously difficult concept to define. Here trust and control are examined and aligned. This would appear to have very little relevance to the current study, except in that it connects with some of the underlying factors (e.g. opportunistic behaviour, or the lack of it) that, from a transaction-cost theoretical perspective, drive *ex-post* transaction costs in construction projects.

—— **(2001)** Dissolving the conceptual barriers to teamwork: reflections on the objectives of institutional economics and social psychology. *University of Melbourne, Department of Management Working Paper in Human Resources Management, Employee Relations and Organizational Studies*, **6**, 1-10.
Abstract: Teamwork forms the basis for innumerable approaches to workplace reorganization and its popularity appears to continue unabated. This is despite the predictions of institutional economics and social psychology that teamwork is an intrinsically inefficient form of organization. This paper examines the objections to teamwork raised by these two disciples, focusing on the problems of "free-riding" or "social loafing" respectively, both of which stem from the exercise of self-interested utility maximizing behaviour in group situations. The paper shows that changes in workplace technology and organization mean that free riding or social loafing are now less likely to occur under conditions of teamwork. In the light of these observations, the implications for further research into teamwork are assessed.
Notes: A refreshing antidote to the bland and uncritical lip service paid to "teamwork" by many "post-Latham" and "post-Egan" commentators.

Shash, A A (1993) Factors considered in tendering decisions by top UK contractors. *Construction Management and Economics*, **11**(2), 111-18.
Abstract: Bid decisions by contractors are complex due to the uncertainty about many factors affecting their outcomes. A questionnaire survey was used to identify 55 factors characterizing the bid decision-making process. The questionnaire was mailed in August 1990, to 300 top contractors in the UK. The results indicate that several factors are considered equally important for the bid/no bid and mark-up decisions. Other factors are seen to have considerable importance for one decision but not for another. The need for work, the number of competitors tendering, and the amount of experience on such projects are identified as the top three factors that affect a contractor's decision to bid for a project. The degree of difficulty, the risk involving owing to the nature of the work, and the current work load are the highest ranked factors affecting mark-up size decisions.
Notes: The paper aims to identify factors influencing a contractor's decision to bid and the mark-up decision. The work is primarily about bidding theory and bidding models, rather than costs associated with bidding.

Shen, L and Song, W (1998) Competitive tendering practice in Chinese construction. *Journal of Construction Engineering and Management, ASCE*, **124**(2), 155-61.
Abstract: In line with the economy reform program, competitive tendering methods have been introduced to the Chinese construction industry to supplement and gradually replace the past assignment system for the procurement of construction projects. The major objectives of applying the competitive methods are to improve the effectiveness of construction investment and to develop the Chinese construction market toward the international procurement practice. This paper examines the development of competitive tendering practices in the Chinese construction industry, particularly in the more developed

regions. It presents the characteristics of the tendering practice within the special environment where the national economy system is being reformed from a purely planned system to a market-oriented system. Outstanding problems in the development of tendering practices have also been investigated. A constructive survey and an in-depth discussion from Chinese academic institutions and governmental departments are used to support the analysis. The study provides a strong indication of the potential development of a competitive tendering system in China.
Notes: Recent history of the development of tendering practices in China, and the results of a small survey about tendering practices. No assessment of the costs of tendering is made.

Shoesmith, D R (1996) A study of the management and procurement of building services work. *Construction Management and Economics*, **14**(2), 93-101.
Abstract: The performance of nominated and domestic sub-contractors in building services in the Hong Kong construction industry is analysed. The variables influencing overall project performance are mapped and tested for the factors that are involved in the selection of sub-contractors, which in turn are likely to influence project performance. There were two objectives: first to investigate the relationship between project management actions and the achievement of reduced costs and completing on time; second, to explore the most appropriate procurement method for building services for particular types of project. To achieve the first objective, the investigation used systems and contingency theory to view managerial actions, and it was hypothesized that the way that the procurement form was managed would determine the project's performance. Moreover, the nature and structure of the temporary multi-organization (TMO) would determine the most appropriate form for optimum performance. In short, appropriate organization management leads to higher performance of building services sub-contractors. The second objective was addressed by considering the roles and responsibilities afforded to the specialist contractor. The methodology for the research was a case study format that enabled the problems experienced in each project to be reviewed. Six case studies of high-rise commercial buildings have been carried out. Although the sample was small, it did provide sufficient data to test the methodology. However the small sample did make it difficult statistically to test the data with confidence. The managerial actions that have been observed arose from a cross-sectional study of data collected by structured interview and a questionnaire that used a standard scoring system. Data on the profile of the sub-contractors undertaking the work were also gathered and management actions of the sub-contractor and project performances were appraised.
Notes: As with many of the papers on construction procurement, this work assesses the options from the standpoint of their purported benefits, rather than their costs.

Siddiqui, W A (1996) Novation: and its comparison with common forms of building procurement. *In:* Harlow, P (Ed), *Construction Papers, No. 60*, Ascot: Chartered Institute of Building.
Abstract: In this paper those types of project for which novation is appropriate will be identified and an assessment made in terms of its ability to satisfy client's criteria for success of their project i.e. time, cost, quality and function/performance. The essential factors that differentiate novation from other project delivery systems will be highlighted and the legal background investigated. The significance of risk transfer to the contractor and the risks still remaining with the client will be examined. The analysis made will enable clients to evaluate the advantages and disadvantages of novation compared to other

procurement options and, thereby, provide a rational basis for selecting that which best suits the characteristics of the project.
Notes: Novation is described and discussed in relation to its applicability as a procurement method, concentrating particularly on the aspect of risk.

Simpson, B (1995) CCT for engineering and technical services: an appraisal. *Proceedings of the Institution of Civil Engineers-Civil Engineering*, **108**(1), 28-32.
Abstract: In just over a year, nearly two thirds of local government construction activity in English metropolitan districts and London boroughs will be subject to compulsory competitive tendering (CCT). Now firmly established for manual services, the government is planning to extend CCT to all non-manual municipal services including engineering. The move is not without controversy. This paper reviews current perceptions and highlights the uncertainty felt by both public and private sector engineers alike over the future of British municipal engineering.
Notes: Now largely superseded because local authorities are no longer subject to CCT. The main thrust is political and concerned with survival of LAs' own direct organizations. Mentions the "inefficiencies and expenses" in competitive tendering (p.30) and that "consultants spend a great deal of time preparing bids for contracts" (p.30).

Skitmore, M and Masden, S E (1988) Which procurement system? Towards a universal procurement selection technique. *Construction Management and Economics*, **6**(1), 71-89.
Abstract: Two approaches are described which aid the selection of the most appropriate procurement arrangements for a building project. The first is a multi-attribute technique based on the National Economic Development Office procurement path decision chart. A small study is described in which the utility factors were weighted by averaging the score of five "experts" for three hypothetical building projects. A concordance analysis is used to provide some evidence of any abnormal data sources. When applied to the study data, one of the experts was seen to be a typical. The second approach is by means of discriminant function. The analysis also found the quality criteria to have no significant impact on the decision process. Both approaches provided identical and intuitively correct answers in the study described. Some concluding remarks are made on the potential of discriminant analysis for the future research and development in procurement selection techniques.
Notes: This is in the same *genre* as many of the papers on construction procurement, whose aim is either to re-classify procurement methods, align them to particular client needs, or measure their purported benefits. This type of work rarely refers to the relative cost of procuring by different methods.

Skitmore, M R and Mills, A (1999) A needs based methodology for classifying construction clients and selecting contractors: comment. *Construction Management and Economics*, **17**(1), 5-7.
Abstract: This note is a comment on Chinyio, E A, Olomolaiye, P O, Kometa, S T and Harris, F.C (1998) A needs based methodology for classifying construction clients and selecting contractors, *Construction Management and Economics*, 16(1), 91-98, which describes research aimed at classifying clients by their needs rather than by the traditional public/private/developer approach. The paper also proposes a new method of selecting contractors by matching clients' needs to contractors' ability to satisfy them. The note offers constructive criticism of some aspects of the analysis.

Notes: Comment on Chinyio *et al.* (1998) highlighting logical and analytical problems with the approach and its application.

Skitmore, M R and Wilcock, J (1994) Estimating processes of smaller builders. *Construction Management and Economics*, **12**(2), 139-54.
Abstract: The paper describes a study of the way in which smaller builders price items in bills of quantities for competitive tender. A series of interviews revealed some marked differences between normal practice and literature-based prescriptions. An experiment was conducted in which eight practising builders' estimators were separately presented with a representative sample of 36 bill of quantities items taken from groundwork, *in situ* concrete work and masonry sections. The estimators stated the method they would normally use to price each item, their "normal" price rate and their highest/lowest price rate. The results showed that only half the items would be priced by the prescribed "detailed" method, the remainder being priced mainly by "experience". Analysis by work section, item rate, item quantity, item total, item labour content, contribution to the total of the bill, the standard deviation of the inter-estimator intra-item rates and totals and their coefficients of variations, skewness and kurtosis indicated that the item total was the main factor determining the rating method used, although this varied in importance between work sections. An intra-estimator intra-item analysis of pricing variability generally confirmed the assumption of a constant coefficient of variation.
Notes: Micro-analysis of the estimating practices of small-sized contractors.

Skoyles, E R (1982) Waste and the estimator. *In:* Harlow, P (Ed), *Technical Information Service, No. 15*. Ascot: Chartered Institute of Building.
Abstract: Not all the materials delivered to construction sites are used for the purpose for which they were ordered, and builders frequently use more materials for which they receive payment. While it is on site that waste occurs, it is not always the result of actions there. Designers of materials, plant and buildings, quantity surveyors and even clients, can all be responsible for losses over which the builder has little control. This paper looks at the types of waste and ways in which the estimator can tackle this waste.
Notes: Developing strategies for more effective waste allowances to be incorporated into estimating practice.

Smith, S (1997) Shell are sure about partnering. *Contract Journal*, **389**(6136), July 30, 18-19.
Abstract: In place of the traditional approach of tendering out to multiple contractors and the adversarial relations that often come with it, Shell has entered into partnership with a single firm for their £350m facelift scheme.
Notes: One of a number of similar articles from the UK press of the late-1990s which either advocate the benefits or point to the uptake of partnering.

Srinivasan, R and Harris, F C (1991) Lane rental contracting. *Construction Management and Economics*, **9**(2), 151-55.
Abstract: The principal lane rental systems operated in the UK are described, and the merits and disadvantages of each, with issues relating to contractual arrangements, management style, contract finish times, working conditions and arrangements.
Notes: "Lane rental" is an innovative contractual approach aimed at incentivizing the timely completion of road projects.

Suraya, I (1997) *A feasibility study in quantifying transaction costs in construction procurement routes in the UK: the case of general contracting and integrated design-and-build*, Unpublished MSc Thesis, Department of Construction Economics and Management, Bartlett School, University College London.

Abstract: The objective of this feasibility study is to illustrate and quantify transaction costs in general contracting and integrated design-and-build procurement routes. Total construction costs consist of production costs and transaction costs. The imperative issue is whether transaction costs can be quantified and subsequently reduced in any of the two procurement routes/governance structures. The study uses background literature of Williamson's theory of incomplete contracting and Stinchcombe's ideas of hierarchy through contract. Governance properties and transaction properties of the construction industry were taken from the ideas of Campagnac, Lin and Winch. Examples of transaction costs are preparing Bills of Quantities, multiplicated estimating effort by tenderers, simultaneous contract management by the consultants and dispute resolution. All transaction items were analysed by the author in her interviews with experienced clients in the UK construction industry. The authors find that clients do not have the actual cost data of transaction costs. They pay the total construction costs but do not have the breakdown between production and transaction costs. This is due to the fact that they are not the party who creates transaction costs but are the ones paying for it. However, the cost of one of the control actors to a client was evident in this study. Furthermore, this study found that clients have taken into consideration transactional properties in the alternative governance structures when given incentives *ex-ante* and *ex-post* to contactors.

Notes: Largely theoretical, but importantly points to the areas of transaction cost "waste" that can accompany an uncritical approach to procurement. The study points out that clients "do not have the actual cost data of transaction costs", underlining the need for research into these costs.

Svensson, R (2001) Success determinants when tendering for international consulting projects. *International Journal of the Economics of Business*, **8**(1), 101-22.

Abstract: A unique database on individual proposals is used to analyse competition among consulting firms (CFs) for international projects. CFs, which sell services based on human capital, focus on developing countries when operating abroad and, thereby, are highly dependent on development agencies (DAs). The DAs have strict tender rules and claim that skill and experience are the most important factors when proposals are evaluated. Both economic theory and the results of the estimations suggest, however, that long-term relationships (LTRs) between the CFs and the clients are at least as important as traditional skill and experience factors. The LTRs are here measured by means of information about whether the CF has previously worked for the client (repeat purchases) or has visited the client. The results indicate that the client in some cases has pre-decided which CF to select. The client invites several CFs to compete for the tender anyway, either because he is forced to do so by the financier, or because he wants to subject an old supplier to competitive pressure. As the tender rules do not seem to be followed, a policy implication would be that the DAs can skip, or at least relax, their strict tender rules, or strengthen the sanctions associated with violations of the rules.

Notes: This paper is about the factors that affect the choice of consulting firms for international projects based on a study of Swedish firms. It was found that long-term relationships were at least as important as traditional skill and experience factors. The interest of the paper for the purposes of our research is that it points at the marketing which consulting firms have to do to obtain work. Clearly, there is a need to assess the costs. They include: (i) Skill education and experience of employees (this is necessary to do good work so not attributable to need to market); (ii) Previous assignments of whole firm and reputation (this is what it is but marketing the data is important); (iii) Long term relationships (one cannot create overnight but there is a cost of keeping in touch); (iv) A permanent or representative office in the host country (a major cost). The paper considers these factors for overseas work but most are in some way applicable in the UK. Most would not vary with the type of procurement.

Swaffield, L M and Pasquire, C L (1999) Examination of relationships between building form and function, and the cost of mechanical and electrical services. *Construction Management and Economics*, **17**(4), 483-92.
Abstract: Relationships between building function, building form and mechanical and electrical services cost, are examined, including the collection of data, and transformation work to enable analysis. Relationships are identified between building form parameters, e.g. perimeter of external walls, gross floor area, storey heights, percentage of glazing, and the mechanical and electrical services costs for buildings of different functions (commercial, industrial and residential). There are relationships between cost of the mechanical and electrical services installations and some building form descriptors, but the particular descriptors and the strength of the relationships vary according to the function of the building.
Notes: The keyword "tender cost" appears to mean price at time of tender, as they make clear in the conclusion that by cost they mean price. There is no assessment of any tendering costs, only of the things that appear to influence pricing levels for M&E service.

Swan, W, Cooper, R, McDermott, P and Wood, G (2001) A review of social network analysis for the IMI trust in construction project. *In:* Akintoye, A (Ed), *17th Annual ARCOM Conference*, 5-7 September 2001, University of Salford. Association of Researchers in Construction Management, Vol. 1, 59-67
Abstract: The issue of trust has been raised as a "gatekeeper" to problems of improving construction procurement (Latham 1994). The IMI Trust in Construction Project is an attempt to evaluate the levels of trust between individuals working together on construction projects. The project is currently undertaking pilot case studies applying Social Network Analysis. This approach has already been demonstrated as useful in reflecting the structure of construction teams. The Trust in Construction approach seeks to expand this to include cognitive and behavioural data based on the ideas of a Trust Inventory. The first stage of the Trust in Construction project is to develop networks built on objective measures, based on questionnaires and documentation, to evaluate structure and information flow. The data collection methodologies are presented as a central factor in evaluating relationships and, ultimately, trust. Social Network Analysis gives a number of analytical approaches to attempt to understand the data collected. The data allows the analysis of structural, transactional and linking concepts. Concepts such as distance and centrality can show roles and participation within the network providing a basis for further analysis. Cliques, clusters or cores can shows how individuals or groups of individuals operate within the context of the whole construction project.
Notes: A theoretical discourse about "trust". As with much of what is written about trust this has little relevance to the current study, except in that it connects with elements of transaction-cost theory such as *ex-post* opportunistic behaviour.

Tah, J H M, Thorpe, A and McCaffer, R (1994) A survey of indirect cost estimating in practice. *Construction Management and Economics*, **12**(1), 31-6.

Abstract: A survey was carried out into practices and attitudes in seven firms towards the quantification and allocation of general overheads, risk contingencies and profit in a tender. The survey indicates that the methods used are highly subjective and are based on past experience. Quantitative methods involving statistics and probability, even though advocated, are rarely used. This suggests that future methods adopted in a computerized estimating environment should reflect the subjective nature of the process and should be simple enough to be applied.

Notes: Although this work examines approaches to the estimating of indirect construction costs, there is no assessment of the costs of resources used during the tendering or estimating stages.

Thompson, R F and McDermott, P (2001) The management of architects within architectural businesses. *In:* Duncan, J (Ed), *CIB World Building Congress*, 2-6 April 2001, Wellington, New Zealand. CIB8.

Abstract: The abandonment by the RIBA in the late 1980s under pressure from the Monopolies and Mergers Commission of mandatory fee scales and the other apparatus of a professional monopoly was seen in the early 1990s as the major factor in an RIBA Study, which was expected to fundamentally change the way architectural businesses were to carry out their trade. The study went on to predict that architects should relinquish certain cherished but obsolete beliefs. A need for strategic thinking was identified by the Study as an important factor in improving the viability of architectural businesses in the early 1990s, and specialization the route to future stability. Regarding architectural businesses, the following paradox within architectural firms has been identified: "Not only is it difficult to anticipate and control workflow and cash flow, the process also involves managing creative professionals who are culturally resistant to being managed. Many architects find the idea of formal planning and adhering to a fixed strategy impractical. This is paradoxical, since architects spend their working lives developing concepts and then detailing plans for their implementation". A qualitative model is developed, based on systems theory, which enables deeper research into architects' businesses. This new model defines a way of comparing the social efficiency between firms and particularly the beliefs or perceptions of the architects that work in these businesses. In general social efficiency entails personal goal attainment on the part of the members at all levels in an organization, and this includes involvement, satisfaction, participation and other variables associated with intrinsic motives and psychological rewards. A comparison of social efficiency is considered here as a more effective technique than the previous models, which are based on a marketing perspective or single measures of profitability. The research through a series of case studies will look at traditional, multidisciplinary, commercial and named architectural businesses. This qualitative approach will examine the following propositions, architects businesses are inherently unstable; architectural businesses that superposition the market place are the most socially efficient; creative architects are difficult to manage. These propositions will be validated through a process of contrast comparison and replication.

Notes: This is of interest to those interested in the costs of tendering and wining work, especially in terms of how consultants win work, in that it investigates the difficulties regularly encountered by one of the main participants in the many construction projects: the architect. There is very little in the literature on how architects incur costs in the running of their own practices, or on how they understand their own cost basis.

Tiong, R L K (1995) Risks and guarantees in build-operate-transfer (BOT) tender. *Journal of Construction Engineering and Management, ASCE,* **121**(2), 183-7.
Abstract: The build-operate-transfer (BOT) concept is being used increasingly by governments across a number of infrastructure sectors in their drive to privatize major projects. Governments see BOT schemes as a method of financing the construction of urgently needed infrastructure projects without direct sovereign guarantee of the loans and with all the technical and financial risks being borne by the private promoter. This paper is concerned with the issues of risks to be retained by the promoter and the guarantees to be offered by the government in the selection process of a BOT tender. It is critical for the promoter to understand that the ability to retain risks and offer guarantees does provide the competitive advantage in being awarded the concession.
Notes: A study of the relationship between success in winning BOT tenders and the allocation of risks and the use of guarantees. This paper does not address the quantification of resources that might be used during tendering.

Tiong, R L K and Alum, J (1997) Distinctive winning elements in BOT tender. *Engineering, Construction and Architectural Management,* **4**(2), 83-94.
Abstract: The Build-Operate-Transfer (BOT) model of development of privatized infrastructure projects is implemented through the award of a concession to a private sector consortium which will finance, build and operate the facility. In a competitive BOT tender, the selection of the successful consortium does not depend on the lowest tolls offered by the tenderer. Rather, it is dependent on the ability of the promoter to provide the most competitive package of distinctive winning elements in its proposal during the final negotiations. The promoter must fully understand the government's needs and concerns and be able to address them through the right package of the winning elements. In this paper, these elements are developed from sub-factors of the critical success factors of technical solution advantage, financial package differentiation and differentiation in guarantees.
Notes: Considers the applicability of the BOT procurement model, and therefore of only limited interest.

—— (1997) Final negotiation in competitive BOT tender. *Journal of Construction Engineering and Management, ASCE,* **123**(1), 6-10.
Abstract: The final negotiation in the competitive tender of a build, operate, and transfer (BOT) infrastructure project is concentrated on the financial and contractual elements associated with the concession agreement and financial package. This paper shows how there are several elements, in particular those related to tolls, that are regarded by governments and promoters as both important and difficult to negotiate during the final negotiation. The competitive tendering and negotiation model is therefore a useful process model for the government and promoters involved in a BOT tender to have a common understanding and to reach a successful conclusion at the final negotiation.
Notes: Research into the final negotiation stages of BOT projects shows that Tiong's process model for competitive tendering and negotiation in BOT projects helps the parties to understand the difficulties in reaching agreement in such projects.

Tookey, J E, Bowen, P A and Hardcastle, C (2002) Construction procurement and management in the UK and South Africa: A comparative study. *In:* Uwakweh, B O and Minkarah, I A (Eds), *The 10th International Symposium of the W65 Commission on*

Organization and Management of Construction, 9-13 September 2002, University of Cincinnati, USA. CRC Press, New York, Vol. 1, 476-96.

Abstract: Recent case study based research into procurement in the Central Belt area of Scotland in the UK has generated a number of interesting findings associated with management on construction projects. The main finding has been that procurement has now evolved into a state best described as "post-modern". Formal procurement systems no longer seem to exist in any recognisable form; instead an infinite variety of procurement systems are generated as clients and professional advisors "pick and mix" desirable qualities of various systems to make bespoke solutions to their perceived situational requirements. These findings in some regards closely match, and occasionally diverge from the results of a wide-ranging survey conducted in the environment of the South African construction industry. While the methodologies adopted by the two studies differed markedly, the similarity in results implies significant movement in the nature and form of procurement and therefore management around the world. This paper therefore compares and contrasts the findings of two very different studies in disparate nations, in order to address the issues of similarity and difference and identify common themes associated with "post-modern procurement" internationally.

Notes: A further attempt to re-classify "modern" procurement.

Tookey, J E, Murray, M, Hardcastle, C and Langford, D (2001) Construction procurement routes: re-defining the contours of construction procurement. *Engineering, Construction and Architectural Management*, **8**(1), 20-30.

Abstract: There are several different types of procurement route available for clients to choose from. Each different type of procurement (traditional, design and build, management, etc) has its own proponents and inherent strengths and weaknesses. Selection of optimal procurement systems is difficult, because even experienced clients cannot know all the potential benefits or risks for each system. Procurement is, therefore, a succession of calculated risks. Industry and academia have focused research on reducing procurement risk through better procurement-system selection methods. Current research considers procurement as a set of rationalistic decisions within a closed environment, aiming to produce generic, prescriptive rules for clients and advisers to use to select the best procurement route for their project. This paper seeks to identify whether prescriptive procurement guidance was adhered to on a set of case study projects. It was found that clients usually selected appropriate procurement systems, and where an inappropriate system was selected, alterations were made in contract form to incorporate aspects of the best procurement route.

Notes: A risk-based approach to procurement and the use of the contract form to adapt procurement solutions.

Tucker, S N and Ambrose, M D (1998) Innovation and evaluation in process improvement. *In:* Hughes, W P (Ed), *14th Annual ARCOM Conference*, September 9-11, University of Reading, UK. Association of Researchers in Construction and Management, Vol. 2, 349-58.

Abstract: Perceived problems of inefficiency in the building process have resulted in development of a range of innovative procurement strategies to overcome problems. Many of the individual strategies have attempted to solve at least one of the inefficiencies in the process, with some more successful than others. An apparently innovative procurement system was examined in an effort to understand the difficulties in re-engineering the procurement process. The project highlighted several problems in trying to be innovative in the design development phase of a project, in particular whether the desired features can be identified and assessed before implementing a new procurement system. The aim of a methodology for assessing a procurement system is to provide a procedure to evaluate the appropriateness of a particular procurement system (or systems) for a specific project and the objectives of the client. A range of factors has been identified as describing the requirements of a procurement system and thus influencing the decision on which procurement system to use. A re-engineered procurement process must be measured against the existing systems and must rate more highly if it is to qualify as an improvement. The approach described here incorporates a number of factors through an interaction matrix, which determines their relative contributions to the success of a project. This interaction matrix is used to combine the strengths of the factors into a single value by which the possible procurement systems can be ranked taking into the account the client objectives and the characteristics of the project.

Notes: As with the majority of such speculative papers on procurement, the "perceived problems of inefficiency" in current systems (which the authors here refer to) are never quantified. It is the need to avoid the unquestioning acceptance of such views that initiated the present study.

Uher, T E (1996) Cost estimating practices in Australian construction. *Engineering, Construction and Architectural Management*, **3**(1/2), 83-95.

Abstract: The aim of the paper is to examine attitudes of general contractors operating in the Sydney region to the potential use of probability estimating and databases in cost estimating. A sample of 10 large general contractors with a turnover over $100m was selected for the study, which took place in 1993. Responses of the contractors to a standard questionnaire were obtained using face to face interviews. The research described in this paper confirmed the popularity of traditional single value estimating and highlighted the lack of use of probability cost estimating by the general contractors surveyed. The limited availability of client-prepared bills of quantities for tendering has neither diminished their popularity among bidding contractors nor increased the use of elemental cost planning. Although databases are generally available, subjective judgements of estimators are of greater value in cost estimating. The research has concluded that a change in the estimating paradigm towards probability cost estimating, and the use of databases, are unlikely to occur in the near future.

Notes: Although primarily concerned with contractors' approaches to estimating, rather than the costs of the approaches, the conclusions are interesting and useful. Clearly, estimators rely heavily on subjective assessments and any attempts to get them to change their approach will need to focus on cultural and attitudinal considerations.

Uher, T E and Runeson, G (1984) Pre-tender and post-tender negotiations in Australia. *Construction Management and Economics*, **2**(3), 185-92.

Abstract: One of the main features of the Australian building industry is the high level of sub-contracting of building works. A survey of 43 sub-contractors is reported, regarding various aspects of the relationship between sub-contractors and general contractors. In particular, the practice of "bid shopping" is highlighted, in which the general contractor in various ways attempts to reduce the sub-contractor's price below that of the tender. Generally the sub-contractors were strongly against bid shopping attempts to tie with general contractors and negotiations in general. However, the strength of the responses was

determined by the size of the firms. The larger firms were more often open to negotiations and deals than the smaller firms. The overall benefits to the general contractors who "shop" are doubtful, most sub-contractors adjust their mark up by up to 20% to allow for such negotiations.
Notes: The work provides evidence of tendering malpractice in sub-contractor procurement.

Vidogah, W and Ndekugri, I E (1998) Improving the management of claims on construction contracts: consultants' perspective. *Construction Management and Economics*, **16**(3), 363-72.
Abstract: There is tremendous scope for improving claims management practice. This research comprised a postal questionnaire survey of contractors, project owners, architects, quantity surveyors and engineers, case studies on actual claims situations on projects, and structured interviews with consultants and contractors. It is based mainly on consultants' views, although contractors' views are brought in occasionally for corroboration and clarification. The main findings are that: (i) claims management is still performed in an ad hoc manner; (ii) contractors' management information systems are ill-designed to support claims; (iii) the products of basic good management practice, such as diaries, timesheets, and programmes, often are inadequate in content even if available; and (iv) some aspects of claims are impossible to quantify with precision even with the best information available at reasonable cost. Main remedial measures suggested include: (a) greater emphasis on the quality of claims management practice and information systems during evaluation of tenders; (b) agreeing figures usually in contention as terms of contracts; (c) implementation of electronic document management systems; and (d) stricter contractual provisions on the quality of programmes, timesheets and content of claims.
Notes: Survey of 54 respondents about claims management practice. Nothing about quantifying the work involved in claiming, but mention is made about the inadequacy of contractors' management information systems.

Wåhlström, O (1991) Simplified tender documents, giving an unambiguous representation of the finished building. *Building Research and Information*, **19**(5), 311-14.
Abstract: A comparison of two ways of tendering and procurement methods in Sweden. It was widely stipulated that Swedish contractors had to invest a considerable amount of time and work when submitting tenders (which included complete and detailed drawings and descriptions), particularly for reasonably smaller projects for which they had virtually no chance of influencing the method of production.
Notes: Tendering and procurement methods invariably influence the total cost of a project. The amount of work and time involved in producing tenders is acknowledged, but the authors do not quantify the resources involved.

Walker, A and Chau, K W (1999) The relationship between construction project management theory and transaction cost economics. *Engineering, Construction and Architectural Management*, **6** (2), 166-76.
Abstract: The process of managing the design and construction of a project may be analysed using project management theory based on a contingency approach. The analysis provided by this approach, while useful for understanding the interaction of the parts of the system, the functions of project management and the effectiveness of the organization structure, may be limited by not incorporating an economic explanation of how a project organization structure is chosen. The transaction cost approach to the study of economic

organization may provide a theoretical basis for such an explanation. An understanding of transaction cost economizing is central to the study of organizations as it determines whether functions are provided by the market or by hierarchy. The relationship between these two powerful approaches is explored, in terms of explaining the structuring and management of project organizations on behalf of clients and to explain the benefits of combining these approaches in furthering construction project management theory.
Notes: A suggestion to use the transaction cost approach to add to the understanding of construction activity.

Walker, D H T (1994) Procurement systems and construction time performance. *In:* Rowlinson, S (Ed), *CIB W92 Procurement Systems Symposium*, 4-7 December 1994, Hong Kong. The Department of Surveying, Hong Kong University, Vol. 1, 343-52.
Abstract: During 1993 a detailed study was undertaken of 33 construction projects built in Melbourne over the last five years. The objective of the study was to gain an understanding of why some buildings are constructed faster than others by identifying risk factors, how builders coped with them, and how they structured their organization and management resources to cope with identified risks. Results revealed that contract type does not significantly affect speed of construction and that several client related factors proved more significant, particularly how well clients relate to the project team. These results pose an interesting insight into the nature of the client/project team relationship and throw some light onto conclusions drawn by others that a non-traditional form of procurement achieves better construction time performance (CTP) results than that of a traditional approach.
Notes: A further example of work that examines the applicability of procurement systems, i.e. their benefits, but stops short of investigating their costs.

Wang, M T and Wu, T S (2000) Cyberspace tendering system, and electronic procurement issues. *International Journal of Computer Integrated Design and Construction*, **2**(2), 134-41.
Abstract: Many government agencies around the world have been establishing their tendering information systems via the Internet to slash lead-times and paperwork in recent years. Although these cyberspace systems can efficiently provide us with some tendering and awarding information, how to implement these systems to automatically complete a construction project procurement process still leaves many reengineering issues to be tackled. Such significant issues include legal problems, adaptation of information technologies, and changes of hardware and software. The objective of this paper is to present a design of a cyberspace tendering and bidding system for improving the construction industry's procurement process. This paper first introduces the importance of construction procurement, followed by a detailed discussion of current practices in the procurement procedure. Some primary problems occurring in the procedure are also described. In this study, a design of a cyberspace tendering system is proposed to solve or avoid these problems. Additionally, major reengineering issues related to the design are also discussed in this paper. Although the proposed cyberspace system still needs to be improved in some areas, it can provide both owners and contractors with a variety of benefits, including more timely updates, faster retrieval of data, better information quality, and shorter response times.
Notes: If information costs are a substantial component of transaction costs, technological advances that reduce the costs of information should have an important impact. Cyberspace tendering systems offer precisely that possibility, and are therefore of general interest.

Wang, S Q, Tiong, R L K, Ting, S K, Chew, D and Ashley, D (1998) Evaluation and competitive tendering of BOT power plant project in China. *Journal of Construction Engineering and Management, ASCE,* **124**(4), 333-41.
Abstract: The tremendous economic growth in China has resulted in an immense demand for basic infrastructure like roads, ports, and power generation facilities. There are thus many investment opportunities for foreign investors. At the same time, some innovations have also been introduced in procurement practices. For example, the build-operate-transfer (BOT) scheme, open competitive bidding, and fully foreign ownership were adopted at the end of 1996, when the concession of the first state-approved BOT project, the Laibin B power plant project in Guangxi Province, was awarded. It is important, therefore, for foreign investors to understand the current regulations, approval procedures, and evaluation and competitive tendering of privatized infrastructure projects by the government-approved BOT project agents in China, in order to secure concession contracts and manage the associated project risks well. This paper discusses these innovations, the roles of the build-operate-transfer project agents, and the evaluation and competitive tendering of these projects in China based on the Laibin B power plant project. The lessons for investing in future similar BOT projects in China are also drawn.
Notes: Advice and guidance for those wishing to tender for BOT work in China, including details of the approval procedures. A useful summary of the development of tendering in China. There is a clear perception in China that competitive tendering increases the quality and efficiency of contractors' performance. Costs of tendering are not considered.

Weston, D C and Gibson, E G (1993) Partnering- project performance in US army corps of engineers. *Journal of Management in Engineering, ASCE,* **9**(4), 410-25.
Abstract: Partnering in engineering and construction usually involves an agreement between an owner and a contractor to work together for an extended period of time, over several consecutive contracts. Because of legal regulations, the US Army Corps of Engineers is unable to establish long-term partnering relationships, but has been successful in implementing partnering on a project-by-project basis. This paper presents an overview of the methodology that is used to set up public-sector agreements. It presents the extent of the Corps of Engineers' partnering program in domestic districts, including the status of projects both completed and ongoing. Data from recently completed projects are presented that indicate significant cost savings and schedule reductions for construction projects executed using partnering arrangements, and interview data with partnering project participants are given that tend to support the premise that partnering is a viable contract administration alternative for public-sector projects. Finally, conclusions and recommendations concerning partnering on Corps of Engineers' projects are presented.
Notes: An overview and study of partnering in context. The "significant cost savings" that are referred to are related to production costs, and the transaction-cost aspect is neglected.

Williamson, O E (1975) *Markets and hierarchies: analysis and antitrust implications: a study in the economics of internal organization.* New York: Free Press.
Abstract: This book is concerned with the organization of economic activity within and between markets and hierarchies. An "organizational failures framework" is proposed and repeatedly employed in an attempt to assess the efficacy of completing related sets of transactions across a market or within a firm.
Notes: Williamson's work is fundamental to the development of transaction cost theory. It is almost entirely theoretical, with no empirical testing; it is rejected by many economists.

However, it is probably true to say that it represents one of the most seductive economic approaches for non-economists, organizational theorists and the like.

—— **(1981)** The economics of organization and the transaction cost approach. *American Journal of Sociology*, **87**(3), 548-77.
Abstract: The transaction cost approach to the study of economic organization regards the transaction as the basic unit of analysis and holds that an understanding of transaction cost economizing is central to the study of organizations. Applications of this approach require that transactions be dimensionalized and that alternative governance structures be described. Economizing is accomplished by assigning transactions to governance structures in a discriminating way. The approach applies both to the determination of efficient boundaries, as between firms and markets, and to the organization of internal transactions, including the design of employment relations. The approach is compared and contrasted with selected parts of the organization theory literature.
Notes: Williamson's "efficient boundaries" conclusions are probably of less interest to construction academics than some of his other work, however, there is some relevance to the "make or buy" decisions that surround the choice between retaining or outsourcing work.

—— **(1981)** Contract analysis: the transaction cost approach. *In:* Burrows, P and Veljanovski, C G (Eds), *The Economic Approach to Law*. London: Butterworth, 39-60.
Abstract: Identifies three dimensions for describing transactions, chief among which is the extent of transaction-specific investment, others being uncertainty and the frequency with which transactions recur. A transaction cost approach is used for the study of contracting, focusing on commercial contracting. Alternative economic approaches to the study of contracting are briefly sketched. Ian MacNeil's 3-way classification of contract is summarized. The essential transaction cost concepts and dimensions needed for matching governance structures with MacNeil's contractual relations are developed. The differential match of governance structures with commercial contracting are then accomplished. Other applications are briefly sketched.
Notes: The identification of the drivers of transactions costs is probably the most interesting (to a construction management academic) of Williamson's contributions to economic theory. There is an obvious connection (explored in the text) with MacNeil's exposition of "relational contracting".

—— **(1985)** *The economic institutions of capitalism: firms, markets, and relational contracting*. New York: The Free Press.
Abstract: The book successively sets out the rudiments of transaction cost economics, applies the basic arguments to a series of economic institutions over which there has been widespread disagreement or puzzlement, and develops public policy ramifications.
Notes: Extends transaction cost theory into public policy.

Wilson, O D and Sharpe, K (1988) Tenders and estimates: a probabilistic model. *Construction Management and Economics*, **6**(3), 225-45.
Abstract: Problems faced by building project procurement agencies have been given only minor attention in the literature. When setting the rules of the tendering competition it is desirable that there be an appropriate balance between reasonable competition and a reasonable cost of competition. This paper examines some 410 projects obtained by a large

state government instrumentally and proposes a model to describe the procurement system. It is probable that the lump sum competitive price in advance system of procurement is unsuitable for small renovation-type works. Larger projects having Bills of Quantities performed very well and satisfactory results can be obtained by inviting four tenders per project. Some suggestions are made for improving the quality of the professional estimate.
Notes: Interesting for its large sample, the paper follows the well-trodden path of re-categorizing procurement systems.

Wilson, O D, Sharpe, K and Kenley, R (1987) Estimates given and tenders received: a comparison. *Construction Management and Economics*, **5**(3), 211-26.
Abstract: Selection of the contractor and the tendering method constitute vital steps in the project procurement process. When budget approvals are based upon professional estimates, an excessive discrepancy between estimate and tender may lead to disruption of construction and budget programmes. This paper examines factors that influence the success of the tendering process within the constraints of a budgetary approval system. Three factors were found to be important in influencing the re-submit rate; namely the number of tenders, whether or not there was a bill of quantities, and whether the project value was less or greater than AU$50,000. The procedure used for analysis was logistic regression applied to grouped and ungrouped data.
Notes: The need to re-submit tenders, where they fall outside client budgets, is an additional cost that sometimes attaches to this type of process.

Winch, G (1989) The construction firm and the construction project: a transaction cost approach. *Construction Management and Economics*, **7**, 331-45.
Abstract: Three of the influential perspectives for analysing construction management are reviewed: socio-technical systems; organization and environment; and project management. It is suggested that in spite of their considerable usefulness, they contain no framework for analysing the inevitable differences in interest between the different firms who are members of the project coalition. An alternative approach is then presented – the transaction cost approach – which, it is suggested, does allow these differences to be analysed. In condition, the dynamics of the contracting system are assessed in terms of the contradiction between construction firms' responses to the uncertainties inherent in the project, and those deriving from the contracting system itself.
Notes: An interesting introduction to the potential for applying a transaction cost approach to construction projects. Analytical rather than empirically-based, this paper makes no attempt to quantify project procurement costs.

Wong, C H, Holt, G D and Cooper, P A (2000) Lowest price or value? Investigation of UK construction clients' tender selection process. *Construction Management and Economics*, **18**(7), 767-74.
Abstract: There is a growing urge for a shift from "lowest-price wins" to "multi-criteria selection" practices in the contractor selection process. The rationale is to achieve best value (for money) for the client. Earlier investigations have found that the tender price (i.e. capital cost) still dominates the final selection decision despite increased emphases on the need for contractor selection based on "value". This paper provides insights into the evaluation of contractors' attributes, particularly for project-specific criteria (PSC), that is, criteria against which tendering contractors may be considered. The importance attached by clients to the "lowest price wins" philosophy is also reported. The perceived importance of

PSC (i.e. their influence on final selection choice) is determined through a structured questionnaire survey of UK construction clients. The results show an increasing use of PSC. "Lowest-price" is not now necessarily the client's principal selection criterion, but rather, the realization that cost has to be tempered with evaluation of PSC in any attempt to identify value for money.
Notes: Based on a summary of a range of papers that are listed rather than reviewed, 37 criteria are listed that might be used for selecting contractors on a basis other than lowest price. A small postal survey of 86 clients sought views on the relative importance of these 37 project-specific criteria. Various statistical manipulations are presented. Not surprisingly, they find that there is a move away from lowest bid wins. Most of the refs in Wong *et al.* are about how to select, rather than the cost or consequences of the decision, just as the paper is.

Youjie, L and Qiang, Z (1997) Building economics research in PRC: a review. *Construction Management and Economics*, **15**(5), 421-8.
Abstract: Research in China into building economics is currently focused on the transition from a centrally planned economy to a market economy. Particular attention is being paid to the theoretical and practical issues arising from the economic reforms. Outside China, there is little known about the various organizations and individuals involved. Their work is funded through the Ministry of Construction and the China National Science Foundation. It is difficult to get commercial sponsorship for this kind of research. Research has focused upon such issues as the role of the construction industry in the national economy, the use of competitive tendering, price formation, urban housing, structuring of the industry and of firms, project management, expert systems and management information systems. There have been some notable successes in terms of building economics researchers influencing the way in which the construction industry is dealt with in China. Future research will be aimed at converting state enterprises to true companies and devising the means to deregulate the pricing of built facilities and construction services without disrupting the market.
Notes: Examination and discussion of the role of construction economists in the transition to a free market. A mention of costs: "In view of the fact that estimating the construction costs of a project is time-consuming, many people in China have been seeking less time-consuming methods of estimating than the traditional manual methods based upon the norms and the standard schedule of unit rates issued by the government agencies in charge". But the real purpose of the paper is to argue the case for building economics research, rather than for any particular methods of estimating.

Zarkada-Fraser, A (2000) A classification of factors influencing participating in collusive tendering agreements. *Journal of Business Ethics*, **23**(3), 269-82.
Abstract: The morality of tendering practices is an issue of economic and social significance, especially when large government contracts are involved. Criticisms are mostly concentrated around collusive tendering: illegal agreements between tenderers that result in seemingly competitive bids, price fixing or market distribution schemes that circumvent the spirit of free competition and defraud clients. Although collusion has been identified as an endemic malaise of tendering, its behavioural and moral dimensions have not been systematically studied before. The paper addresses this knowledge gap and describes part of an exploratory empirical investigation of the decision-making patterns utilized by estimators in Australia's largest construction organizations.
Notes: This paper is about the factors influencing collusion in tendering.

Zhu, T (2000) Hold-ups, simple contracts and information acquisition. *Journal of Economic Behavior and Organization*, **42**(4), 549-60.

Abstract: In a typical procurement set up, several recent papers have shown that when complete contracting is not possible, simple, non-contingent contracts may suffice to solve the under-investment problem. This paper points out that a non-contingent contract offer such as a fixed-price contract may induce the seller to acquire information on the future course of costs and only to accept the offer if the cost is low. It is shown that sometimes the buyer prefers to wait and buy on the spot market than to offer a long-term contract. When the seller rejects a contract offer or the buyer chooses not to make one, the seller will not make efficient investments because he expects to be held up on the spot market.

Notes: This paper explores some interesting ideas of the effects of asymmetry in the information of parties to a contract before and after signing. The relevance to the present study would arise in considering enforcement costs with various types of contractual arrangements.